The Butterflies of Egypt

This study is dedicated to

Edward P. Wiltshire, CBE

for his work on Middle East Lepidoptera
including the first comprehensive review of
Egyptian butterflies

THE BUTTERFLIES OF EGYPT

TORBEN
B. LARSEN

APOLLO BOOKS / THE AMERICAN UNIVERSITY IN CAIRO PRESS

Edited by Leif Lyneborg.
Dust jacket by Jan Haugum.
Illustrations by the author and British Museum (Natural History).

Published outside Egypt by
Apollo Books,
Lundbyvej 36,
DK-5700 Svendborg,
Denmark.

Published in Egypt by
The American University in Cairo Press,
113 Sharia Kasr el Aini,
Cairo, Egypt.

ISBN 87-88757-14-5 (Apollo Books).
ISBN 977 424 218 1 (The American University in Cairo Press),
Dar el Kutub No. 8854/89.

Printed in Denmark by Vinderup Bogtrykkeri A/S.

Contents

The 8 colour plates are arranged after page 79.

Abstract

This book attempts to summarise the current knowledge of the butterflies (Lepidoptera; Rhopalocera) of Egypt and to place them in ecological and biogeographical perspective. The fauna is not rich, encompassing 58 species, since climatic conditions are not conducive to butterfly survival. Egypt is positioned as a very arid environment interposed between the more mesic parts of the eastern Mediterranean and North Africa, and bordering on the Afrotropical Region in the south. Much of the Egyptian fauna is therefore found only on the eastern, western and southern borders of the country. Levels of endemicity are very low. Details are given as far as possible on the distribution, habits, ecology and biology of the taxa involved. The distribution within Egypt and the zoogeographical patterns are analysed.

Key words:
Lepidoptera; Rhopalocera - Egypt - Mediterranean - zoogeography.

Introduction

The aim of this book is to review all that is known about the butterflies of Egypt, and to place this information in biogeographical perspective. Such a review is long overdue. The last account dates back forty years to the well-known paper by Wiltshire (1948), itself the first major review since the publication of Andres & Seitz (1923). Since then additional information has come to light on the Egyptian fauna, but more importantly, the knowledge of neighbouring faunae has increased dramatically. This permits a much clearer biogeographical interpretation of the limited but interesting fauna. Finally, there have also been many changes in taxonomy and nomenclature during the past forty years.

Since starting work on Middle Eastern butterflies twenty years ago I was aware of the desirability of an updated review of Egyptian butterflies. They form, in effect, the southwestern frontier for the Levant fauna, an eastern filter for the South Mediterranean fauna, and a tenuous bridge between the Mediterranean and Africa. Despite paying brief visits to Cairo, Alexandria and Luxor on more than a dozen occasions between 1970 and 1977, I did not wish to publish such a review without personal acquaintance with all the varied habitat types that I had not had the opportunity of visiting on previous occasions.

Thanks to the generosity of the Carlsberg Foundation of Denmark it was possible to visit Egypt for six weeks from late March to early May of 1987, covering some 8000 km by car. Compared to my trip to Yemen (Larsen, 1982a) there were obviously fewer original discoveries; the fauna of Egypt was already much better known and in arid climates new discoveries are much more of a hit and miss affair than in areas with more stable climate. The style of the paper is, therefore, somewhat less based on my personal research than was the case in previous papers and books. Nonetheless, the trip was essential to the writing of this review which completes for the time being my researches into Middle Eastern butterflies. Without the support of the Carlsberg Foundation for my trips to Jordan in 1977 and to the Yemen Arab Republic in 1981, as well as the present trip, it is doubtful whether the book would have been written.

Country background

This account covers the geographical area of the modern Arab Republic of Egypt, situated in the northeastern corner of the African continent, but including the Sinai Peninsula in Asia. Modern Egypt more or less encompasses the same territory that was administered by the more dynamic Pharaonic dynasties as much as five thousand years ago, though outlying areas such as Nubia, the Sinai and the Siwa oasis were not always under centralised administration. Egypt is probably the geographical term which for longest has been used in more or less the same way. It is, therefore, perhaps fitting that the oldest depiction of a recognisable butterfly species should be that of a *Danaus chrysippus* in Luxor (Larsen, 1977c; figured in colour by Larsen, 1984a).

Egypt covers a land area of about one million square kilometres. The country straddles the eremic zone (see Larsen, 1984c for a full discussion of this concept), roughly matching the term Saharo-Sindian as defined by Eig (1931) or the African Indian floristic zone by Good (1977). About 95% of the country is desert, the central western desert constituting one of the most extreme arid desert habitats in the world. Rainfall is low in the deserts but, more importantly, it is very erratic and there may be several consecutive years with no rain. Only the coastal strip from Sallum to Alexandria, the Gebel Elba in the extreme southeast, and the higher parts of the South Sinai mountains receives higher and less erratic rainfall. In all three cases this is strongly reflected in the floral and faunal composition. There are a number of oases in the western desert, the largest being the Siwa oasis near the Libyan border. They have a vigorous vegetation by desert standards, but not one that is rich in species. Most of Egypt's landmass is below 500 m, which further limits potential diversity. The mountains of the Red Sea coast are extremely dry with limited vegetation cover, not least compared to the Hejaz mountains on the other side of the Red Sea. The high mountains of the southern Sinai are much more interesting with clearly defined phytogeographical zonation.

The most prominent geographical feature of Egypt, of course, is the river Nile and its delta. Silt carrying waters from the East African mountains have deposited soil in the valley during the annual inundations, creating a continuous 'oasis' habitat from the Sudan to the Mediterranean. This continuous habitat is now cut by Lake Nasser, created by the Aswan High Dam, and silt has not been deposited since the original Aswan Dam was built. More than 5000 years of human activity has totally transformed the Nile Valley. Indeed, anthropogenic influence has been more radical than almost anywhere else. Accelerating population growth this century resulted in the irrigation of even marginal lands. At the turn of the century when the first Egyptian butterfly records were being published the total population was almost certainly below ten million; it is now well over forty. The effects on habitats have been drastic. Not only have the large mammals (lion, hippopotamus, oryx, baboon, etc) become extinct in historical times, but papyrus (*Cyperus papyrus*) on which they were so often lovingly painted in pharaonic times has also disappeared.

Although Egypt has been an arid country since the Miocene, the oldest relevant period for considering current butterfly distribution, there has been a marked process of desiccation since the post-pluvial climatic optimum after the last glaciation. Probably even the pharaohs saw a richer

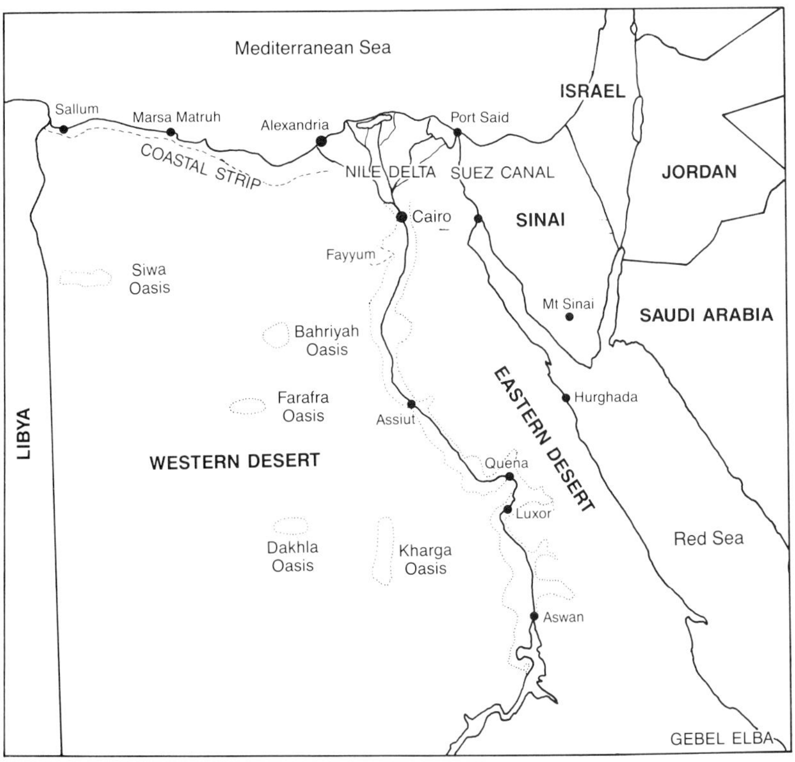

Figure 1. Map of Egypt and the ecological zones used in this book.

desert vegetation than the one we see to-day. More rain and a slightly warmer climate would have allowed the African savanna elements to penetrate into the deserts from the south, and probably Palaearctic elements would have found better conditions on the coast and in the Sinai than they do to-day. Little trace of these periods is found among the extant butterflies, except for the two endemic Sinai species and the isolated presence of *Pseudotergumia pisidice*.

Ecological Sub-Divisions of Egypt

For the purposes of this paper I have divided Egypt into seven main ecological regions, brief descriptions of which are given below with the intention of helping readers of the systematic part who are not familiar with the country. These regions will also be used for analysing the distribution of the Egyptian fauna. The regions are relatively crude. Botanists would doubtless include more subdivisions, emphasising the gradual changes in the composition of the desert flora as one moves south towards the tropics. A brief sketch on the sub-divisions in the Sinai will be included in the rele-

Figure 2. Spring flowers in the Mediterranean coastal strip.

vant section. However, given the low number of butterflies in Egypt any attempt at further precision would probably serve to confuse rather than to illuminate.

THE COASTAL STRIP stretching from Alexandria to the Libyan border receives more, and more consistent, rainfall than the rest of Egypt, supporting a strip of vegetation that has Mediterranean or Irano-Turanian affinities (Fig. 2). It was most developed in the Mariut west of Alexandria though this area is now very degraded. Despite its limited size it contains three butterflies found in no other part of Egypt in the form of *Tomares ballus mareoticus* Graves, *Pseudophilotes abencerragus coloniarum* Turati, and *Carcharodus stauderi ramses* Reverdin. The first and the last of these have subspecies endemic to the zone. The more widespread Palaearctic butterflies also occur in most of the coastal strip. Essentially the coastal strip acts as an eastern extension of Mediterranean elements centered on North Africa and the Iberian Peninsula which otherwise do not extend further east than Cyrenaica. None of these elements has been able to get a foothold in the irrigated delta. Graves (1916) believed that investigation of the area between Alexandria, Mersa Matruh and Sallum would yield a number of further species. My own experience between Alexandria and Mersa Matruh does not support such optimism. Cultivation, irrigation, overgrazing and construction are destroying much of the best Mediterranean habitat. Benyamini (pers. comm.) believes it is quite possible that the three species mentioned above are now extinct in their Egyptian subspecies.

2 THE WESTERN DESERT (also known as the Libyan desert) (Plate 1:3) constitutes the largest part of Egypt. It is a barren expanse of stony or sandy wastes, rarely exceeding an altitude of 300 m with large areas below sea level. With no mountains, hardly any rain, extreme heat in summer and occasional frost in winter the western desert represents one of the most inhospitable habitats for butterflies anywhere in the world and very few have been recorded except on the fringes of oases. Even the oases mainly have a fauna that is characterised by migratory species. The scattered vegetation of the western desert is both much less diverse and much less relevant to butterflies than that of the depressions in the eastern desert, consisting almost exclusively of Saharo-Sindian elements. The extreme dessication and impoverishment of the western desert is of quite recent origin, but the area had no refuges like the Central Saharan massifs to leave testimony of more mesic conditions. Even in pharaonic times grazing and

hunting would still have been possible on a much larger scale than to-day.

Between the Nile and the Libyan border lie the oases of Dakhla, Kharga, Farafra and Bahriyah. Their original vegetation would have been essentially Saharo-Sindian with somewhat more diversity than the present desert plants. They would have been clustered around stands of *Tamarix*, *Phoenix* and *Hyphaene*, with tall grasses fringing water sources. With the advent of cultivation and irrigation the nature of the oases has been drastically changed. Crops and their associated weeds are now dominant, allowing some dispersive Palaearctic butterflies to establish themselves in places they could not have colonised before the advent of man. The oasis of Siwa (Fig. 3) on the Libyan border is larger and more compact than the other three, as well as more isolated. It contains one species of butterfly (*Zizina otis sangra* Moore) otherwise only found in the Oriental Region. However, the butterfly fauna of all the oases is very limited, thirteen having been recorded from Siwa which is the most diverse.

3 THE LOWER NILE AND THE DELTA house the bulk of Egypt's population and have been radically transformed by human activity. Originally, the delta consisted of extensive marshes with natural vegetation in

Figure 3.The Siwa Oasis in the Western Desert.

those parts that were covered by the annual inundations of the Nile. Increasingly sophisticated agricultural methods, better water management and drainage have now left only traces of the original natural vegetation. It is not an exaggeration to say that 95% of the total biomass consists of crops, introduced weeds, and deliberately planted trees (Plate 1:1). Even the watering canals are clogged up with exotic weeds. Of the total biomass at any given time more than half is accounted for by plants that are not native to Egypt. Figs, grapes, dates, barley, forms of durra and certain vegetables were grown from native stock during the Old Kingdom; wheat, watermelon, peaches and many fruits and vegetables were added as trade and conquest widened Egypt's choices; however, many of Egypt's dominant crops are recent introductions, i.e. citrus, cotton, tomatoes, potatoes and sugar cane. Exotic alfalfa is now one of the most important animal feeds as well. Basically the Delta is one giant oasis fashioned by human activity. With one or two exceptions the resident butterflies are those that utilise cultivated plants, their associated weeds, or those remnants of the original vegetation that survive because they are useful. The artificial nature of the habitat is underscored by the fact that there is normally no transition zone to the desert. The border is as if drawn with a ruler (Fig. 6).

Figure 4. The abrupt transition from desert to the Nile Valley near Saqqara.

4 THE UPPER NILE VALLEY, taken to begin at Assiut, is very similar to the delta in most respects and the whole Nile Valley could be treated as one region from Rosetta and Damietta in the north to Aswan in the south. There are at least two differences, however, that are of importance to butterflies. First, the climate becomes progressively hotter, allowing the survival of migrant tropical species all year round. Second, rather more of the natural vegetation remains, especially where the Nile passes through narrow defiles where irrigation is not possible, cf. Figs 5 & 6 and Plate 1:2. In such areas the Afrotropical *Chilades eleusis* Demaison may be found, together with the tropical Lycaenidae that feed on *Acacia*.

5 THE EASTERN DESERT lies between the Nile and its delta, and the Red Sea and the Suez Canal. The northern part of the desert, such as between Cairo and Ismailia or Suez, has a series of small wadis with much more profuse and more diverse vegetation than is seen in the western desert (Plate 2:1). Some Irano-Turanian elements mix with the Saharo-Sindian, creating well developed plant communities consisting of both perennials and annuals. The main flowering season is in April and May,

Figure 5. In places the Nile Valley vegetation is almost non-existent, e. g. near Aswan.

which is also the flight period for the most interesting desert butterflies. It is sometimes possible to find all the food plants in the same depression, though it is rare to find all the butterflies together: *Pithuranthos tortuosus* for *Papilio saharae*; *Zilla spinosa* for *Pontia glauconome*; *Astragalus forsskali* for *Agrodiaetus loewii*; *Kickxia aegyptiaca* for *Melitaea deserticola*; and desert *Convolvulus* for *Spialia doris*. The butterflies are much scarcer than the plants and their geographical range is less extensive.

Further south, between the Nile and the Red Sea, rainfall appears to decline and the plant associations become more impoverished. This is despite the presence of mountains which at times exceed 2000 metres. However, the higher peaks are usually tall, jagged and isolated, with little vegetation and even high up apparently mostly of the Saharo-Sindian type. Very few butterflies are resident in this area, perhaps only *Pontia glauconome*, though the common migrants utilise temporary food resources. The ability of *Lampides boeticus* to exploit any suitable plant in an otherwise totally hostile habitat is particularly impressive. It is possible that pockets of interesting biotopes exist in the wastes of the eastern desert, but I have found no information on this and personally found no interesting sites.

Figure 6. Agricultral land near Qena – the Dum Palm (*Hyphaene thebaica*) is almost the only vestige of original vegetation.

6 THE GEBEL ELBA lies in the extreme southeastern corner of Egypt in territory under joint administration between Egypt and Sudan. This mountain massif has a much more profuse and a much more tropical vegetation than the mountains north of them, such as at Hurgada. The main reason is that these mountains are subject to a monsoon climatic regime. In fact, the Gebel Elba represents the northernmost extension of the dry savannas of Sudan. Several Afrotropical butterflies have been recorded from here that are found nowhere else in Egypt, with *Charaxes hansali* Felder taking pride of place. It proved impossible for me to obtain permission to visit the area, but the butterflies recorded and descriptions would indicate that it is not dissimilar to the facing mountains of the Asir (see Pittaway, 1986). This is substantiated by the special flora of the area (Täckholm, 1974). There is no doubt that several, perhaps many, species not yet recorded from Egypt would be found on a trip, but each new discovery would only serve to underline that the area is the northernmost enclave of the Sudanese fauna whose relevance to the basic fauna of Egypt is marginal.

Figure 7. The Santa Katarina Monastery on Mt. Sinai; both vegetation and butterfly fauna are very rich.

7 THE SINAI is biogeographically somewhat complex (cf. Fig. 7 and Plate 2:2). The northern coast is a Saharo-Sindian area with strong Mediterranean representation. It is a poor habitat for butterflies, containing mostly the common, mobile species. One or two Palaearctic species just manage to penetrate the area (*Polyommatus icarus*, *Carcharodus alceae*). The western coast, on the Gulf of Suez is mostly an alluvial plain with little vegetation and fewer butterflies. The coastal area on the Gulf of Aqaba is also very dry, but the limited vegetation has a much stronger tropical flavour than the Red Sea coast in Egypt proper. The area can be regarded as a northern extension of the conditions prevailing in the Hejaz. There are not many butterflies, but they include a number of tropical species such as *Colotis phisadia* Godart and *Anthene amarah* Guérin-Méneville, as well as the two species of *Azanus*.

The Central Sinai is the best developed Saharo-Sindian habitat in Egypt and contains all the eremic species, but since there is much Irano-Turanian relict vegetation, a number of Palaearctic species also occur. It is an extension of the Negev. It is the high mountains of the southern Sinai, though, that are the most interesting. The flora is mainly Irano-Turanian, but there are a large number of Mediterranean plants and a high degree of endemism. The vegetation is much richer than elsewhere in Egypt and a visit during the main flowering season in late April and early May is a delight. The area contains disjunct populations of several Palaearctic butterflies as well as two outstanding endemics in the form of *Satyrium jebelia* and *Pseudophilotes sinaicus*, both described by Nakamura as recently as 1975. Almost forty species of butterflies have been recorded from the southern Sinai.

History of Butterfly Research in Egypt

Egypt never evoked the same enthusiastic response among butterfly researchers as did Turkey, Lebanon and Israel. The fauna was perceived as being limited and uninteresting which, while partly true, is not really the case. There was real romance and mystery concerning the Sinai, but that area was difficult of access, and anything near a comprehensive set of data had to await the Israeli occupation from following the 1967 war. It was, in fact, only in 1948 that Wiltshire's justly celebrated faunal study produced anything resembling a full account of the butterflies of Egypt as a whole. Since then, during the past forty years very little has been published.

The first Egyptian butterflies in press were a few from the Sinai collected on the Hemprich & Ehrenberg expedition to Ethiopia, Arabia and the Sinai in the beginning of the 19th century. The two explorers did a marvellous job, and their rich butterfly material was published by Klug (Klug & Ehrenberg, 1829-1845) with the most beautiful of hand coloured plates. Klug's work was extremely meticulous and exact. *Pseudotergumia pisidice* and *Pontia glauconome* were expressly published as from Sinai Monte. So was the taxon now known as *Colotis danae eupompe*, but that was probably due to a labelling error.

In 1870 Francis Walker published a list of the material collected by J. K. Lord esq. in Egypt and the Sinai. This included a number of Egyptian firsts, but Walker described his material in such a haphazard manner that one of his obituaries began: 'More than twenty years too late for his scientific reputation, and after having done an amount of injury to entomology almost inconceivable in its immensity, Francis Walker has passed from amongst us' (Larsen, 1979). The paper was therefore largely ignored; a status report on the taxa contained is given by Nakamura (1977). Some of his names survive as valid in this review though most were found to be synonyms.

By the end of the 19th century quite a few records from Egypt had found their way into the many butterfly publications being published at the time, but there were still none dealing specifically with the Egyptian fauna. It seems that the first such paper was that of Bethune-Baker (1894) dealing with a small collection received from Alexandria. Demaison (1895) published a short list of butterflies from the Aswan/Abu Simbel area collected by his brother some ten years earlier.

The beginning of the 20th century saw a flush of publications on the Egyptian butterflies. In 1902 there was a botanical expedition to the Sinai, where Kneucker and Guyot collected butterflies, recording a total of seven species from the peninsula (Kneucker, 1903). A more detailed review by Rebel & Daub (1909) brought the total to fourteen, including previous records. One of the most interesting Kneucker records was *Euchloe falloui* Allard, a record disbelieved by Wiltshire (1948), and a species not recorded again till collected in the Sinai seventy years later by Nakamura & Benyamini (1973). However, it was not till the concise review of Benyamini (1984) that anything resembling a full review of the Sinai butterflies became available.

Around the turn of the century the Societé Entomologique d'Egypte was founded, based on the rich bourgeoisie of the trade and cotton community that made Alexandria what it was at the time. It was later to be named in honour of King Fouad I who had a personal interest in butterflies. Parts of his collection can still be seen at the Manyal Palace in Cairo. Innes-Bey (1909, 1909a) listed the publications relevant also to Egyptian butterflies, and a number of Egyptians and expatriates began to publish information on the local fauna. Details will be found in the list of references. It was Andres (1912) who published the first real consolidated list of Egyptian butterflies. P. P. Graves, later to become involved in the running of the T. E. Lawrence operation and who was knowledgeable on the butterflies of the Middle East, was less than impressed with the effort, and published his own list in 1916, partly as a rebuttal. In 1923 Andres & Seitz updated the earlier list and the Sinai material without adding much that was new, but for the first time attempting to place it in a wider biogeographical and ecological context.

Hayward (1925ff) published rather detailed observations on the butterflies of Aswan, and Williams (1929) studied the detailed phenology at Giza of four of the more common species. In 1935 the Armstrong College of Cambridge University arranged a lengthy expedition to the Siwa oasis; its butterflies were treated by Gabriel & Corbet (1949).

The main contribution to Egyptian butterflies to date was Wiltshire's careful, annotated catalogue, summarising earlier experience in the light of his extensive experience in the neighbouring countries. But he was also the first to report in detail on the large, well labelled collection formed over many years by A. Alfieri, a veritable gold mine of information, containing many species new to Egypt and *Iolana alfierii* as new to science. Wiltshire's list contained 53 genuine species. Few have been added since though the ranges of several have been extended.

Since Wiltshire (1948), very little has been published except for the Sinai paper of Benyamini previously alluded to. Roell (1959) spent the month of April (ca. 1957) in Egypt, but he did not make any significant discoveries. Some biological information on the pest species of butterflies has appeared in the *Bulletin de la Societé Entomologique de l'Egypte*, the most interesting being one on *Deudorix livia* by Awadallah et al. (1970). My own visits have resulted in the extension of the known ranges of a number of species, but not in any species new to the country. During six weeks in 1987, I paid special attention to the Siwa Oasis, the coastal strip and the Red Sea coast in the hope of finding things of special interest. I failed to do so and I am convinced that this is because few species remain to be discovered in Egypt (except for the Gebel Elba).

I have gone through the collections of the British Museum (Natural History). Contrary to my expectations they contain very little material from Egypt and I found nothing of special interest. This is in marked contrast to the rich material from the Sudan.

Neighbouring areas

Before commencing on the systematic review of the Egyptian butterflies it is useful to give a few notes on the extent to which neighbouring areas have been investigated in respect of butterflies. There have, over the last fifteen years, been a considerable number of publications which are of use in the interpretation of the Egyptian fauna, and some of them are still not as widely known as they deserve. Reference is not made to earlier papers where the recent ones cited contain adequate bibliographic references.

The Lebanese fauna was recently revised by Larsen (1974), while the Syrian fauna remains practically unknown. It is bound to be a fragmented one, with great contrast between the northern mountains, the Antilebanon, and the arid zone. A closer study should throw light on the connection between the Taurus and Lebanon. Larsen & Nakamura (1983) recently reviewed the Jordanian fauna. Benyamini (1983) has produced a concise and precise list of the Israeli fauna. During 1988/89 he will be publishing a book in Hebrew on the Israeli butterflies, including the fauna of Mt. Hermon and the Sinai. One hopes this will eventually be translated into English. Parker (1983) has given an excellent and interesting account of the Cyprus fauna.

Moving further south, the Sinai fauna was summarised in a concise paper by Benyamini (1984), almost doubling the number of species recorded by Wiltshire (1948).

Arabian butterflies had been relatively unknown till I started a research programme on the peninsula. This has resulted in a detailed treatment of the Oman fauna (Larsen, 1977d; 1980b; Larsen & Larsen, 1980), the fauna of the Yemen Arab Republic (Larsen, 1982a), and finally the Arabian Peninsula as a whole (Larsen, 1983). Parallel to this work, and in constant consultation, Pittaway (1979, 1980) dealt with the butterflies of eastern Arabia and Qatar. He later published a major paper on the butterflies of western Saudi Arabia (Pittaway, 1985) with many interesting field observations.

The eastern flank of Egypt has thus, so to speak, been well covered. This is not true for the Sudan in the south. The only attempt at a comprehensive treatment of the Sudan butterfly fauna is that of Wilson. It is an excellent work on its own terms, but it is out of date and not detailed enough to be of much help to the serious student. Egypt's western neighbour, the Libyan Arab Republic, has been the subject of some research, mainly during the time when it was illegally occupied by Italy. This resulted in a number of papers by count Turati, scattered in the Italian entomological literature, but never fully digested. There is a fair amount of unpublished Libyan material in the collections of the British Museum (Natural History). A review of the Libyan fauna would be highly desirable. Finally, Tunisia was the subject of a monograph by Chneour (1954). It remains a sound piece of work but needs considerable updating.

Systematic List of Egyptian Butterflies

PAPILIONIDAE

Papilioninae

1 *Papilio saharae saharae* Oberthür

Papilio machaon saharae Oberthür, 1879, Etud. ent., IV:68.
Plate 3: 1.

DISTRIBUTION: This species is a vicariant of the common European and Holarctic *Papilio machaon* Linné. It ranges through the Saharan fringes from Morocco to Egypt, the Sinai, the Negev, South Jordan and the Hejaz, a further subspecies occurring in the Asir (Pittaway, 1985) and the Yemen Arab Republic (Larsen, 1980a, 1983). It is essentially allopatric with respect to *P. machaon*. The decision as to whether its should be considered Palaearctic or eremic is a difficult one since it inhabits mesic montane zones in Arabia. I list it as Palaearctic.

STATUS: *P. saharae* is normally considered a subspecies of *P. machaon*, but I have raised it to specific rank on the grounds that: 1) There are constant differences in genitalia; 2) there are constant differences in the shape of the hindwing tornal area; 3) the early stages differ, as well as the larval food plant choice; 4) there is at least some evidence for interspecific sterility in North Africa (Clarke & Sheppard, 1956); and 5) differing choice of habitat which makes it unlikely that hybrids could survive (Larsen, 1983; Clarke & Larsen, 1986). However, it must be said that hybrids between Yemeni *P. saharae rathjensi* Warnecke and Omani *P. machaon muetingi* Seyer were viable, and Clarke opts to treat the two as conspecific. I prefer to separate them as two distinct biospecies and the most recent field workers with experience of the species in nature agree (Benyamini, 1984; Pittaway, 1985; Meerman & Boomsma, 1986). The taxonomic position of the *P. machaon*-complex in North America is also unclear (Hancock, 1983).

There are not many records of this butterfly from Egypt; they include near Sallum, the Siwa oasis, and various localities in the eastern desert between Cairo and Suez. The large size, the black and yellow colour pattern, and the tails on the hindwings make it unmistakable. Sexual dimorphism is slight. It has several broods, though that in spring is the most numerous. It is not too uncommon in the Sinai where it may be found up to the very top of the highest mountains. I have seen it hilltopping round the little chapel at the summit of Gebel Musa (Mt. Sinai) at 2300m. The main food plant is probably *Pithuranthos tortuosus* which is very much more common than the butterfly, but it probably feeds also on *Haplophyllum*. *Pycnocycla tomentosa* is an important food plant in the Hejaz (Pittaway, 1985), but it appears too rare in the Sinai to be significant. Material from North Sinai may pertain to *Papilio machaon* (normally a larger butterfly, with less angular wings, and with a much larger red eye-spot at the anal angle of the hindwings).

PIERIDAE

Pierinae

2 *Artogeia rapae leucosoma* Schawerda

Pieris leucosoma Schawerda, 1905, Verh. zool.-bot. Ges. Wien, 55:516.
Plate 4: 1-2.

DISTRIBUTION: This is a Palaearctic butterfly found practically throughout the region. It has been recorded from Singapore and was accidentally introduced to Australia where it is now widespread, as happened also in North America and Hawaii.

STATUS: Ssp. *leucosoma* is conventionally used for the subspecies of the eastern Mediterranean, to which the phenotype in Egypt conforms. It is, however, not easy to delineate subspecies of *A. rapae* and I would not quarrel strongly with someone allocating the Egyptian population to the nominate subspecies. Kudrna (1974) removed the smaller members of the genus *Pieris* to *Artogeia*. He has since changed his view on this (Kudrna, 1986). I still think that the *P. brassicae* Linné-group is distinctive enough to merit a separate genus.

A. rapae is one of the most common and best known butterflies in Egypt. The light yellow underside of the hindwings is only shared with the much larger (and very rare) *Pieris brassicae*; the other white Pieridae in Egypt all have prominent black or green markings on the underside. The female has two black discal spots on the discal area of the forewings, the male none or only one. The species may be found flying conspicuously in all cultivated areas from the Mediterranean to Aswan. It is one of the very few Palaearctic butterflies to extend so far south. Migrations are known from Europe, but none has been recorded in Egypt. However, the butterfly clearly has strong powers of dispersal and even tiny oasis habitats in the desert are soon colonised. In the Sinai, too, it is chiefly an oasis species. In April 1987 I found it plentiful in the giant, circular, irrigated vineyards on the Cairo-Ismailia road. Curiously, though, there are no records from the Siwa oasis. The adult butterflies are very fond of the flowers of alfalfa (*Medicago sative*). The larvae feed on a variety of Cruciferae and constitute a pest on cultivated cabbages.

3 *Pieris brassicae brassicae* Linné

Papilio brassicae Linné, 1758, Syst. Ent.,10:467.
Plate 3: 5.

DISTRIBUTION: This is a western Palaearctic butterfly, common throughout Europe, European Asia, the Levant, and parts of the Middle East to the Himalaya. It has been accidentally introduced to Chile. As a migrant it also reaches Egypt on occasion.

STATUS: In my Lebanon book (Larsen, 1974) I considered the name *catoleuca* Röber valid for the Levantine populations. I would now agree with Feltwell (1981) that subspecies in the area cannot be defined.

This common European butterfly is definitely not a true member of the Egyptian fauna. Though abundant in Lebanon in most years, it is much less frequent in Israel, and absent from the Sinai. The only Egyptian record in print is a single specimen from the Mariut near Alexandria in vi.1912. Wiltshire (1948) considered this to be a vagrant from Cyrenaica, a theory that I consider plausible given the lack of records from the Sinai. The species is larger than the preceding one, with deeper black markings. Given that *Artogeia rapae* has established itself so well in the cultivated parts of Egypt one might wonder that this species has not done so, being a notorious pest of cabbages. Probably summer temperatures are too high to allow for permanent populations, an easily testable hypothesis.

4 *Pontia daplidice daplidice* Linné

Papilio daplidice Linné, 1758, Syst. Nat., 10:468.
Plate 4: 4-6.

DISTRIBUTION: This migrant Palaearctic species has a vast range from Europe and North Africa, through the Middle East to East Asia. There are populations also in the Oman mountains, in the Yemen and Asir mountains and in the Ethiopian highlands.

STATUS: Morphologically it is doubtful whether any populations in the western Mediterranean could be subspecifically distinguished. However, Geiger & Scholl (1982) through electrophoretical analysis suggested that there were two species involved in the complex. One, the nominate species, in the Canary Islands, the Iberian area to NW Italy, as well as in Israel; and another in Central Europe, Yugoslavia and Greece. I have had no opportunity to make a personal evaluation of the data. The concept of a North African/Iberian taxon and a Central European/Pontomediterranean one is zoogeographically acceptable, but it is difficult to imagine that the level of migratory activity demonstrated by the species could allow for the continued existence of two species in the Mediterranean basin.

This is the first of two rather similar white butterflies, which have a characteristic green pattern on the underside of the hindwings. *P. daplidice* lacks the clear yellow underlining of the veins that is characteristic of *P. glauconome*. Being highly mobile *P. daplidice* can be met with practically anywhere in Egypt, but its main breeding grounds are the cultivated areas, though populations are sometimes met with in the deserts so that both species of *Pontia* are found in the same spot. Normally the species is unpredictable and relatively rare in Egypt proper, though considerably more common in the Sinai except during the winter months. Probably the species is not a permanent resident except in the Sinai though, as with other migrants, it may survive for several years. The larval food plants are species of Resedaceae.

5 *Pontia glauconome* Klug

Pontia glauconome Klug, 1829, Sym. Phys.: text, pl. 7, figs. 18, 19.
Plate 4: 7-9.

DISTRIBUTION: This desert butterfly has an almost prototypical distribution for an eremic species, from Mauritania to the Horn of Africa, the

Sinai, Negev, the Jordanian Desert, all of Arabia (except the wettest mountains), and the arid zones of Iraq to Pakistan and Afghanistan.

STATUS: A number of subspecies have been described for this butterfly, but they do not appear convincing. The Egyptian populations are definitely of the nominate form.

P. glauconome is very similar to the preceding species but may be distinguished by the distinct yellow underlining of the veins of the hindwing underside. In both species the females are somewhat more strongly marked with black than are the males. This butterfly is found in the arid parts of Egypt where it is well distributed though not normally very common. There are records from Siwa, but it should be found elsewhere in the western deserts. It seems to be most common in the eastern desert between Cairo and Suez and in the Sinai. There are records from the Aswan area, and I found it in several places in the Red Sea mountains. It is so specialised to the desert habitats that it is hardly ever recorded in the delta or on the banks of the Nile in the south. In Egypt the species is closely associated with *Zilla spinosa*. *Ochradenus* is acepted in the Sinai; Andres & Seitz (1923) also record it on *Cleome arabica*. Other Cruciferae and Resedaceae may be used on occasion. Contrary to the preceding migrant species, *P. glauconome* is highly sedentary and even small colonies may be found in the same spot year after year. Its main survival tool in the desert areas is an ability to spend at least four years in pupal diapause when necessary, a remarkable feat considering that temperatures vary from below zero to above 50 Celsius.

6 *Euchloe aegyptiaca* Verity

Euchloe belia race *aegyptiaca* Verity, 1911, Rhop. Pal.:337.
Plate 4: 3.

DISTRIBUTION: Defined as a species, the taxon *aegyptiaca* is found from Libya to Egypt, the Sinai, Negev, South Jordan and the Hejaz (Pittaway, 1985). The species was described after a specimen from Wadi Hof near Helwan and may be considered part of the eremic element.

STATUS: The entire *Euchloe ausonia*-complex is in a state of taxonomic flux. The taxonomic and nomenclatural end result is difficult to anticipate. My own impression is that the Egyptian member of the complex will survive as a distinct species under the present name. I consider it an eremic derivative of the species-group.

E. aegyptiaca is readily recognised by the irregular green markings of the hindwing underside, very different from the paler markings of the two preceding species as well as from the more regular bands that characterise the two following species. It is a rather rare butterfly in Egypt, having been taken in the Mariut area near Alexandria, in the eastern desert, and in the Sinai. The main brood is in spring (Feb.-March), but second brood specimens also occur in May, though very rarely. According to Wiltshire (1948) there is also a specimen from Damanhour, a delta oasis locality; I am inclined to think that it is actually from the near-by desert. In the Sinai the larva feeds on *Diplotaxis harra*, but doubtless other desert Cruciferae are also acceptable.

7 ***Euchloe belemia belemia*** **Esper**

Anthocharis belemia Esper, 1805, Schmett. Abb. Beschr., Bd. 1, Suppl.:92.
Plate 4: 10-12.

DISTRIBUTION: *E. belemia* has a most peculiar distribution pattern for a Palaearctic butterfly, apparently centered on the Mediterranean. From the Iberian Peninsula and adjacent North Africa it extends narrowly along the North African coast to Egypt and the Levant, thence to Iraq and then, again narrowly along both coasts of the Gulf to the Musandam Peninsula in Oman and the Pasni area of the Makran coast. There are isolated populations at high levels in the Asir/Yemen mountains and in Ethiopia.

STATUS: A number of subspecies of this butterfly have been described, but with the exception of the isolated population in Ethiopia (ssp. *abyssinica* Riley) they do not appear well founded.

E. belemia is very similar to the *E. falloui* in the markings of the underside of the hindwing which form parallel green bands; on the forewing underside *E. belemia* has the black spot at the end of the cell centered with white while the spot is completely black in *E. falloui*. Sexual dimorphism is slight.

This is a fairly common butterfly in the coastal Mediterranean zone and in the delta, including the Fayyoum oasis. It does not extend much south of Cairo, nor is it found in the deserts or in the Sinai, being replaced by the following species in desert habitats. I have found the species in the large, circular irrigated vineyards on the Cairo-Suez road to which it must be a very recent immigrant. There are two or three broods between December and April, summer being spent in pupal diapause. In consequence of its habitat choice the species is almost never wholly sympatric with *Euchloe falloui* though the two may occur in close proximity. Most of the small desert cruciferous plants have been noted as larval food plants, though locally one species is usually preferred.

8 ***Euchloe falloui falloui* Allard**

Anthocharis falloui Allard, 1867, Ann. Soc. Ent. Fr.: 318.
Plate 4: 13-15.

DISTRIBUTION: This is a classic eremic butterfly, found throughout the Sahara to the Sinai, Negev, Jordanian desert, and the Hejaz, then south to Jeddah and the Riyadh area. It does not extend to the Indian subcontinent.

STATUS: The nominate form is found in Egypt. Several names for Saharan populations are probably invalid. The Riyadh area population, where males invariably have washed out markings on the underside, still seems deserving of subspecific status (ssp. *saudi* Larsen (1983)).

The species is very similar to the the preceding one, under which the distinguishing features are given. This butterfly is widespread in the Sinai, but there are no authenticated specimens from Egypt proper. I feel sure that given its overall range, including Libya, and its normal habitat choice, it must occur somewhere in the deserts. The species is generally quite local in the Sahara. In the Sinai there are two or three broods between November and May, but they are not well defined and may be missing in some years. The species has developed the capacity to spend several years in pupal diapause. I have seen *E. falloui* hilltopping around the chapel on the summit of Gebel Musa at 2300m. According to Benyamini (1984) the preferred larval food plant in the Sinai is *Moricandia sinaica,* with *Zilla, Diplotaxis, Reseda* and *Schouwia* as stand-bys.

9 *Elphinstonia charlonia charlonia* Donzel

Anthocharis charlonia Donzel, 1842, Ann. Soc. ent. France, 11:197.
Plate 5: 5-6.

DISTRIBUTION: This is an eremic species stretching from Mauritania through the Sahara and its fringes to the Sinai, Jordan, Israel, the Hejaz and Central Arabia. It also recurs in various subspecies from Oman to Baluchistan and the Punjab; these will possibly be found to be specifically distinct such as is definitely the case with the Pontomediterranean populations. I suspected (Larsen, 1983) that the species in Oman was limited to the Musandam, but David Bayley informs me he collected specimens in the Ghubra Bowl of the Jabal Akhdar.

STATUS: The nominate subspecies is found throughout the African range as well as in the Sinai, Jordan, Israel, and Arabia. In their monograph of the species Back & Eitschberger (1976) erroneously synonymised the Jordanian ssp. *elisabethae* Hemming with the Pontomediterranean ssp. *penia* Freyer. This mistake was corrected by Larsen (1977a). Larsen & Nakamura (1983) subsequently sank the name *elisabethae* since there are no consistent differences between Jordanian/Israeli and Saharan material. Back (1982) showed that the Pontomediterranean taxon *penia* is specifically distinct, thus removing the biogeographical anomaly of having an eremic species with a wholly Palaearctic subspecies.

This is normally a highly localised butterfly and it may appear in very small populations. The bright yellow ground colour renders it unmistakable. It has well developed diapause mechanisms and one, even all, of its broods may be missing in a given locality for several years. There are very few Egyptian records. A male was collected at Siwa in April, 1935; small numbers were taken early this century in the Mariut steppe near Alexandria; a few have been taken in various parts of the Sinai. The flight period is between December and April. It is probably more widespread and less rare than the few records imply. The larval food plants are mostly species of *Diplotaxis*, but other Cruciferae are also used.

10 *Colotis phisadia phisadia* Godart

Pieris phisadia Godart, 1819, Enc. Méth., 9:132.
Plate 5: 4.

DISTRIBUTION: This Afrotropical species is found in the Sahel zone north of the Equator, including some of the Saharan massifs. It extends south to dry Kenya and north to the Jordan Valley. It is widely distributed in Arabia and extends to NW India in a well defined subspecies (ssp. *protractus* Butler).

STATUS: Egyptian, Jordanian and Arabian specimens all belong to the nominate subspecies from the Sahel.

This representative of the large Afrotropical genus *Colotis* can be recognised through the broad, black margins to the wings and the characteristic salmon ground colour of the forewings. The black markings of the female are more diffuse and the ground colour is variable. Egypt is very much at the northern limits of the species and there are few records. Alfieri caught two at Hurgada in July, 1935; I found neither the butterfly nor the food plant around Hurgada in 1987. It is quite likely that colonies do exist along the Red Sea coast, though they may not be permanent, since the species is migratory with good temporary colonising ability. The Gebel Elba is certain to have permanent populations of this butterfly. It is more firmly established along the east coast of the Sinai, where it is on the wing from April to October/November. I found no trace of it at Dahab despite searching large stands of *Salvadora persica*, the only larval food plant available in Egypt.

11 *Colotis chrysonome chrysonome* Klug

Pontia chrysonome Klug, 1829, Sym. Phys.: text, pl. 7, figs. 9-11.
Plate 5: 1-3.

DISTRIBUTION: This Afrotropical species is found in the African Sahel and Arabia, with an extension to the Dead Sea and Wadi Araba. A few defined subspecies fly in dry parts of Ethiopia and in Somalia.

STATUS: Egyptian material pertains to the nominate form.

This butterfly can hardly be counted as a member of the Egyptian fauna. The single record is a male from Gebel Elba collected by Alfierii in 1928. There is no chance that it will be found elsewhere in Egypt, except perhaps as a stray from the Wadi Araba in the extreme NE Sinai. The golden ground colour distinguishes it from other Egyptian butterflies. The male has a white base to the forewing upperside that is missing in the female. The larval food plants are species of *Maerua* (Capparidaceae).

12 *Colotis danae eupompe* Klug

Pontia eupompe Klug, 1829, Sym. Phys.,: text, pl. 6, figs. 11-14.
Plate 5: 10.

DISTRIBUTION: This is an Afrotropical butterfly, also maintaining a foothold in India, where the nominate subspecies is found. It is common in southwestern Arabia and Dhofar; scattered colonies are found in eastern Oman. It is one of the most widespread butterflies in dry tropical Africa, just reaching Egypt in the Gebel Elba area.

STATUS: Egyptian specimens belong to ssp. *eupompe* which is found in Africa north of the Sahara and in Arabia.

The male of this butterfly is readily recognised through the crimson tips of the forewings. The female is more strongly marked with black and may lack the crimson wing tips. This is another species which has only been collected once in Egypt, again by Alfierii in the Gebel Elba. It is not to be expected elsewhere in the country. There are two male specimens in the British Museum (Natural History) marked 'Egypt, E. H. Armstrong' from the turn of the century. They may, however, be from the Sudan. Walker (1870) records the species from Cairo; this is quite improbable and Walker lists other species from Cairo that do not occur in Egypt. Klug & Ehrenberg (1829), in Klug's description of the taxon *eupompe* expressly stated that the habitat was 'Arabia deserta et Sinai Monte'. Several other authors have cited the species from the Sinai; all these records are restatements of Klug's record. Almost certainly this is an error in Klug's meticulous work for which there is a precedent. In the description of *Precis limnoria* it was expressly stated that the single specimen was from Arabia Felix (Yemen), though it is of obvious Ethiopian extraction (Larsen, 1983). Given the conditions of the Hemprich & Ehrenberg expedition it is hardly surprising that the occasional labelling mistake should have been made. The larval food plants of *C. danae* are species of *Cadaba* and perhaps other Capparidaceae.

13 *Colotus ? liagore* Klug

Plate 5: 11-12.

DISTRIBUTION: *C. liagore* is found in the driest parts of East Africa, through Arabia to the Makran coast of Baluchistan. There are specimens from Pasni on the Makran coast in the British Museum of Natural History, and I have seen another Baluchi specimen in the collection of the Indian Institute of Agricultural Research in Delhi.

STATUS: Wiltshire (1948) reported a specimen of the very similar *Colotis evenina casta* Gerstäcker from the Gebel Elba. This is far to the north of the range of that species, while the presence of *C. liagore* in the Gebel Elba would not really be surprising. The latter species would not have been well known to Wiltshire at the time. Another possibility is that of *Colotis evagore* Klug or *Colotis ephyia* Klug which should occur in southern Egypt, but I do not think Wiltshire could have confused these species with *C. evenina.* However, without having access to the specimen in question all that can be said is that one of these three species also exists in the Gebel Elba.

This species resembles the previous one, but the wing tips are orange rather than crimson. As mentioned above, Wiltshire recorded a single male specimen collected by Alfieri in the Gebel Elba in 1928 and allocated it to *C. evenina.* I think that the very similar *C. liagore* is a more likely possibility. The determination was made at a time when Wiltshire had little experience with Arabian and Sudanese material and limited reference material available. However, at least one species of *Colotis* with orange wing tips does exists in the Gebel Elba.

14 *Madais fausta fausta* Olivier

Papilio fausta Olivier, 1804, Voy. Oth., 4:29.
Plate 5: 9.

DISTRIBUTION: This is an Oriental species with a distribution dissimilar to any other. It is found in two subspecies in India and Sri Lanka, then in the nominate subspecies from Pakistan and southern Iran, through Arabia and Iraq to the Levant and parts of the East African coast. The nominate subspecies, but apparently not the Indian ones, is migratory. Though they normally do not penetrate Africa this migratory capacity must be responsible for the surprising record of a typical specimen in the Ennedi Massif of central Sahara (Bernardi & Gillet, 1962).

STATUS: Egyptian specimens are of the nominate subspecies. The taxon is normally ranged as a species of *Colotis* but deserves a separate genus.

M. fausta is a widespread and sometimes common species in Egypt east of the Nile and in the Sinai, though there are large fluctuations in numbers from year to year. It was first recorded by Manders (1915). The beautiful salmon ground colour and the limited black markings make the species easy to recognise. Females are more strongly marked with black than are the males. It invades Lebanon and Iraq every year to breed in quantity during autumn, but then dies out. As such it is clearly migratory which raises the puzzle as to why it never seems to reach the Nile valley or the delta, much less to cross it. Wiltshire (1948) thinks the Egyptian populations may be non-migratory; there is not sufficient data to pronounce on this, but it is likely that permanent populations exist in Egypt and the Sinai. It is on the wing from March till October. The larval food plants are *Capparis cartilaginea* and *C. spinosa.*

15 *Anaphaeis aurota* Fabricius

Papilio aurota Fabricius, 1793, Ent. Syst., 3(1):197.
Plate 4: 16-17.

DISTRIBUTION: The range of this Palaeotropical migrant covers dry tropical Africa, Arabia and the Indian subcontinent. Migrants reach Lebanon, Israel and Jordan on an irregular basis, occasionally breeding in large quantities.

STATUS: No subspecies should be recognised. I have taken long series in Madagascar in 1980 and 1982 which in my view do not deserve the subspecific status that they have been accorded by some authors.

The clearly defined black veins of the hindwing underside will immediately serve to distinguish this species from other Egyptian Pieridae. It seems to be quite scarce in Egypt where it cannot survive on a permanent basis except, perhaps, in the Gebel Elba. There are scattered records from the Nile Valley, the Dakhla Oasis, and the Sinai, but they are few and far between. There are none from the western desert proper. I saw large numbers near Cairo International Airport in early December, 1970. This is the latest year it has been seen in the eastern Mediterranean (Larsen, 1976). They may have been remnants of a rather strong invasion of this species to the eastern Mediterranean in 1969, which was the last time it was seen in the Levant, including the Sinai. In April, 1987 I found some clumps of *Capparis aphylla* on the Luxor/Aswan road which showed unmistakable signs of damage by the larvae of the species, apparently from last autumn. When the species does move it often does so in large numbers. In Kenya I once recorded a flight of more than four million over a narrow front in just one day (Larsen, 1977b). I have also given an impression of the capacity for rapid population build-up in Oman (Larsen, 1982b). The larval food plants are *Capparis* and *Maerua*.

Coliadinae

16 *Catopsilia florella* Fabricius

Papilio florella Fabricius, 1775, Syst. Ent.:479.
Plate 3: 2-3.

DISTRIBUTION: This Palaeotropical butterfly is found throughout Africa, Arabia and the Indian sub-continent. Migratory specimens make their way to the eastern Mediterranean at irregular intervals. It has recently established permanent populations in the Canary Islands, but normally seems unable to survive winters in the Mediterranean basin.

STATUS: The species will probably be found to be conspecific with the Oriental *C. pyranthe* Linné in which case this name takes priority and the range is extended to cover all of Asia.

This is a large, powerful butterfly, with somewhat angular forewings. The male is bluish-white; the female may be white or yellow. This tropical migrant is a rather regular visitor to Egypt and has been recorded from the Delta, the Nile Valley, the eastern desert, and the Sinai. Sometimes it builds up large populations and can be very common in the centre of Cairo. It is a much less common visitor to Lebanon, Israel and Jordan. There was a fifty year gap between Lebanese records till I saw a female in 1977 (Larsen, 1978b). Benyamini (1984) suggests that a permanent colony exists in the Eilat area and the east coast of the Sinai; I suspect that winters in this area are sometimes too severe for the permanent survival of the species. It is not a permanent resident of Egypt, though it may survive milder winters even in the Cairo area. It should have permanent populations in the Gebel Elba. In the eastern desert and in the Sinai the main food plant is the common *Cassia italica*, but in the cultivated areas several exotic, ornamental *Cassia* are also used.

17 *Colias croceus* Fourcroy

Papilio croceus Fourcroy, 1785, Entomologie, Paris, 2:250.
Plate 5: 7-8.

DISTRIBUTION: This is a western Palaearctic migrant with considerable powers of dispersal. Its headquarters are the Mediterranean basin and the warmer parts of the Middle East. It regularly colonises Europe during the summer, and Arabia and Egypt during winter.

STATUS: The species has had an unstable nomenclatural life. It is generally agreed to-day that *C. croceus* is a western Palaearctic species, replaced by *C. electo* Linné in southwestern Arabia, and *C. fieldii* Ménétries from the Himalaya eastwards. Egyptian specimens in no way differ from European ones.

The males of this butterfly is orange with broad, black margins; the females are usually orange, but there is a much scarcer white form; both female forms have spots of the ground colour in the black margins. It is generally a common butterfly in Egypt. Though migration is important in the life cycle of the species, it is probably a permanent resident of parts of Egypt. The population dynamics are, however, probably strongly dependent on migration. In Egypt the species is found in all cultivated areas, including outlying oases such as Siwa and Dakhla, especially where alfalfa (*Medicago sative*) is grown, since this is a favourite larval food plant. However, many wild leguminous plants are also accepted. Benyamini (1984) records *Tephrosia* and *Astragalus* in the southern Sinai, as well as *Colutea* (pers. comm.).

LYCAENIDAE

Theclinae

18 *Deudorix livia* Klug

Lycaena livia Klug, 1834, Sym. Phys., text, pl. 29, figs. 11,14.
Plate 5: 13-15.

DISTRIBUTION: This is an Afrotropical species widely distributed in dry sub-Saharan Africa, Arabia and as a migrant in the eastern Mediterranean. There is also a single female specimen from southwestern Iran in the collections of the British Museum (Natural History).

STATUS: There are no subspecies of this migratory insect.

D. livia males have a characteristic copper red colour; the female is brown with some blue suffusion. The shape of the hindwings, with its lobe and tails, is distinct from all other Egyptian Lycaenidae. The species is generally a scarce and unpredictable migrant to the eastern Mediterranean, but it is widespread and often common in Egypt, possibly a permanent resident in Upper Egypt. The breeding experiments by Awadallah et al. (1970) certainly indicate that it could permanently survive there. It is generally limited to the cultivated tracts and does not occur in the deserts except as a rare migrant. Normally single specimens are found feeding from the flowers of alfalfa (*Medicago sative*) or *Acacia*. However, just south of Aswan I found very large numbers on the latter in April, 1987. The species is highly polyphagous. It feeds on *Acacia* as do other members of the genus, but also on a bewildering array of other plants, including young dates, pomegranate, and loquat – occasionally achieving pest status.

19 *Satyrium jebelia* Nakamura

Strymonidia jebelia Nakamura, 1979, J. Ent. (B), 44:283.
Plate 6: 4.

DISTRIBUTION: The species is endemic to the high mountains of the southern Sinai.

STATUS: The generic status of the Holarctic species of the group of genera to which *S. jebelia* belongs is of extreme complexity. I follow Clench (1979) in allocating it to *Satyrium*. Kudrna (1986) uses *Nordmannia* for this species group. As pointed out by Nakamura there is no doubt that the species is most closely related to the western Palaearctic *S. spini* Denis & Schiffermüller, but it is very distinct.

A glance at the plate will show that this butterfly cannot be confused with any other Egyptian species. Sexual dimorphism is not pronounced, but the male has a small androconial brand in the cell of the forewing upperside. It is limited to the mountains of the southern Sinai where it flies at levels between 1800 and 2200 metres. It was first collected by Benyamini on Gebel Bab in 1974 and forms a most interesting addition to the Palaearctic fauna. The period of flight is from late May to mid-June in a single brood. In the right place the species is not uncommon, though very localised. The larval food plant is *Rhamnus dispermus*. Full details of this interesting butterfly are given in the original description.

20 *Tomares ballus mareoticus* Graves

Tomares (Thestor) ballus mareoticus Graves, 1918, Entomologist, 51:97.
Plate 6: 7-9.

DISTRIBUTION: This is a Palaearctic species with a West Mediterranean distribution, ranging from the French Riviera via the Iberian peninsula and North Africa to Egypt. It is a characteristic distribution pattern, though the eastern limit of the Mediterranean range varies. Most West Mediterranean species do not reach Egypt; a few extend to the Levant, and *Euchloe belemia* and *Carcharodus stauderi* have extended far beyond that.

STATUS: The consistent presence of red markings on the upperside of the male forewings probably justifies Graves' erection of a specifically Egyptian subspecies. It is however, very close to the Cyrenaican population, later named as ssp. *cyrenaica* Turati. This may turn out to be a synonym of *mareoticus*.

The green underside of both sexes in this butterfly is characteristic; males usually have slight red patches on the forewing upperside while females have much more red. The only records of this interesting butterfly are from the Mariut steppe near Alexandria, with its Mediterranean floristic character. There are two specimens from 'Cairo' in the British Museum of Natural History. They are actually the Mariut specimens from Dekhla mentioned by Bethune-Baker (1894). The species occurred only sparingly in the first decade of this century and I know of no later specimens. I totally failed to find it. The species is univoltine sometime between January and March. It normally flies in short grass, being somewhat inactive. The green underside affords good camouflage. Nothing is known of its food plant in Egypt, but like elsewhere in the Mediterranean it is likely to be *Lotus* which is found in its Egyptian range.

21 *Apharitis acamas acamas* Klug

Lycaena acamas Klug, 1834, Sym. Phys., pl. 40, figs. 7-9.
Plate 6: 1-2.

DISTRIBUTION: This is a prototypical eremic butterfly found in the arid zone from the central Saharan Hoggar Mountains to Egypt, the Levant, Middle East and Arabia, and thence to western Iran, Afghanistan, Pakistan and northwestern India.

STATUS: I have recently been involved in a review of this species (Larsen & Pittaway, 1982). Egyptian material is ranged as nominate ssp. *acamas*. The validity of a separate taxon, ssp. *egyptiaca* Riley (1925) for the populations of Egypt is considered unnecessary. The subspecies of this butterfly are rather more clearly defined than those of the following species.

This is the first of two similar species that are not always easily told apart because of significant individual variation in both size and colour pattern. Both are orange above with some dark brown markings, while the underside have traverse bands that are centered with dull, metallic scales. *A. acamas* is normally a darker orange than is *A. myrmecophila*, with more dark markings above, and with all dark markings above and below arranged in bands that are less macular than in the latter. The species is localised and not normally common. Egyptian records are from the Mokattam Hills near Cairo, various localities in the eastern desert, and the Sinai (according to Benyamini (1984) as high as 1950m). It is certain to occur in the western desert as well. It normally occurs in desert terrain with some vegetation, but can also be found under oasis conditions. While the larva may feed on a number of different plants, it also seems able to survive inside ants' nest, feeding on ant larvae and pupae. Larsen & Pittaway (1982) reviews what is known of its biology.

22 *Apharitis myrmecophila myrmecophila* Dumont

Cigaritis myrmecophila Dumont, 1922, Bull. Soc. ent. Fr.:217.
Plate 6: 3.

DISTRIBUTION: This eremic species is found in a string of ill-defined subspecies from the central Sahara, to Egypt, the Middle East and Arabia, as well as southwestern Iran.

STATUS: Each population of this species seems to have its own phenotype and the variation is not well expressed in traditional subspecific terms. However, the Iranian populations do appear to merit genuine subspecific status as ssp. *farsica* Rose & Schurian (1977).

The species is close to the preceding one, under which the distinguishing characters are reviewed. *A. myrmecophila* is a rare and enigmatic insect which few people have seen in nature. It is normally found in small colonies that are strongly localised and with very low numbers of individuals. It is often found with the desert plant *Calligonum comosum*. Partly because of the extreme localisation, Egyptian records are few and far between, chiefly in the Suez Canal zone and the southern Sinai. It is the only butterfly to maintain permanent populations in the Arabian Rub al Khali desert and is likely to be found in the Egyptian deserts as well. The larvae are probably able to live in ants' nests without feeding on green plants as is the case in the preceding species (Larsen & Pittaway, 1982).

Lycaeninae

23 *Lycaena phlaeas phlaeas* Linné

Papilio phlaeas Linné, 1761, Faun. Suec.:285.
Plate 5: 18.

DISTRIBUTION: This is a Holarctic butterfly that is common throughout the northern temperate zone, with a considerable capacity for pushing its frontiers to the limit of its ecological tolerance. Isolated populations exist in the high mountains of southwestern Arabia and East Africa.

STATUS: In my Lebanon book (Larsen, 1974) I used the subspecific name *timeus* Cramer. I do not now think that a Levantine subspecies can be justified. The populations inhabiting SW Arabia and East Africa are closely allied to each other.

A glance at the plate will show that *L. phlaeas* is an unmistakable butterfly. Sexual dimorphism is slight. There are very few records of this butterfly from Egypt. It was caught in Cairo (Urman Gardens, Giza) in August 1945. There is also a record from Mersa Matruh (Andres, 1930), described as a new form of the species. There are also some records from the Sinai, but Benyamini (1984) feels uncertain that it is a permanent resident. My personal reaction is one of surprise that the species is not well established in the Sinai and the delta, but presumably summer temperatures are too high. The larval food plants, *Rumex*, are common and should not be a limiting factor.

24 *Lycaena thersamon omphale* Klug

Lycaena omphale Klug, 1834, Sym. Phys., pl. 40, figs. 12-14.
Plate 5: 16-17.

DISTRIBUTION: This beautiful little butterfly is found from Italy and Austria to the Balkans in localised colonies, then more generally in the Middle East to Afghanistan, just penetrating Egypt in the Sinai.

STATUS: The subspecific taxonomy and nomenclature of the species has been very confused. I have used the subspecific names *kurdistanica* Riley (Larsen, 1974), following Paulus & Rose (1971), and *omphale* Klug (Larsen & Nakamura, 1983). Schurian & Hofmann (1982) argue that the Levant populations should be regarded as pertaining to the nominate subspecies. I have not been able to reach a final personal decision on the merits of this argument.

The yellow/orange underside of this butterfly, with its profusion of black spots, will immediately serve to distinguish it from other Egyptian species. Above the male is a beautiful, fiery orange; the female is somewhat more subdued and has a profusion of dark brown spots matching those of the underside. This species hardly belongs to the Egyptian fauna, having been collected only twice in the northern parts of the Sinai. Alfieri found it in the Wadi umm Mitla on 8.v.1932; Benyamini at Kadesh Barnea on 3.iii.1978. In Israel and Jordan there are broods between March and October, the later broods normally being larger than the first. I have always found the adults to be very partial to the flowers of *Eryngium creticum*. The larval food plants are mainly *Rumex*, *Polygonum* and *Sarothamnus* (Polygonaceae).

Polyommatinae

25 *Anthene amarah amarah* Guérin-Méneville

Polyommatus amarah Guérin-Méneville, 1849, Lef. Voy. Abyss.:348.
Plate 6: 14-16.

DISTRIBUTION: This Afrotropical butterfly is found throughout Africa, except in the rainforest zones. It extends to Arabia. Records from Aqaba, the Sinai and in the Bombay area are probably due to the dispersive potential of the species.

STATUS: The nominate subspecies is found throughout the range except for a small pocket within the Liberian rainforest.

Both sexes of this butterfly have three little 'hairtails' on the fringe of the hindwings, but as shown on the plate the underside pattern is also distinctive. The male upperside is a metallic lead; the female is brown with a varying degree of lighter markings. The species cannot easily be confused with other Egyptian butterflies. *A. amarah* is hardly a resident of Egypt. Records are from the Gebel Elba and the Sinai, one from South Sinai, others from near Eilat. There are no actual records of migration, but the species is, at the very least, able to extend to the very frontiers of its potential range. I am surprised it has never been found in the Aswan area. The larval food plants are species of *Acacia.*

26 *Lampides boeticus* Linné

Papilio boeticus Linné, 1767, Syst. Nat., 12:789.
Plate 7: 12-14.

DISTRIBUTION: This Palaeotropical migrant is found throughout the old world tropics, extending well into the Palaearctic, though it is usually unable to survive on a permanent basis in a non-tropical environment.

STATUS: No subspecies should be recognised, though several have been proposed.

The underside markings are identical in both sexes; as shown on the plate they are most distinctive. The male is uniformly violet blue above, while the female is dark brown with blue markings on the discs of all four wings. This is probably the most common butterfly in Egypt. It may be found practically anywhere, though numbers fluctuate. Few butterflies have more colonising power than *L. boeticus.* It is not unusual to find that even single leguminous plants in the middle of a vast barren expanse contains larvae of the species (e.g. between Hurgadha and the Nile or in the western deserts). It qualifies for pest status since cultivated beans and peas are high on its list of preference for larval food plants. The larvae feeds inside pods and can render them unmarketable. However, it also feeds on a wide range of wild leguminous plants. *Colutea istria* is a great favourite in the Sinai mountains.

27 *Syntarucus pirithous* Linné

Papilio pirithous Linné, 1767, Syst. Nat.,12:790.
Plate 6: 17-19.

DISTRIBUTION: This is an Afrotropical migrant butterfly found throughout Africa and most of Arabia. There are permanent populations in various parts of the Mediterranean, including Egypt (Larsen, 1986).

STATUS: In older works the names *telicanus* Lang or *plinius* Fabricius are usually used, but all records are of *S. pirithous*. The Egyptian population is often referred to ssp. *egyptiaca* Bethune-Baker (1894) but does not seem to be well enough differentiated to deserve a subspecific name of its own.

The underside pattern of this butterfly is sufficient for identification purposes; the male is blue above, the female is blue with darker markings. *S.pirithous* is one of the most common butterflies in Egypt, especially in the oases, but also in the Mediterranean zone, the eastern desert, and the Sinai. Its abundance in alfalfa fields is sometimes quite remarkable (I have also seen it in masses at the Gezira Club in Cairo). The species is mostly on the wing from March till October though it is possible to find the occasional specimen in winter. It is a somewhat migratory species and may build up populations for some time in areas where permanent survival is not possible. The larval food plants are a wide array of leguminous plants and several plants unusual for members of the Lycaenidae, *Plumbago* being a great favourite. This plant is found in the Gebel Elba area.

28 *Tarucus balkanicus* Freyer

Papilio balkanicus Freyer, 1844, Neu. Beitr. Schmett., 5:63.
Plate 6: 20-22.

DISTRIBUTION: This is an essentially eremic species with a centre of distribution to the north of other members of the genus. It is found from NW India and Afghanistan, through the Middle East to the Balkans and Arabia. It is, however, conspicuously absent from Yemen and the Asir. It recurs in Sudan (I have personally collected it in Kharthoum), thence through the Sahel to Mauritania and the coastal districts of Tunisia and Algeria. It is found in the northern Sinai, but is unaccountably absent from Egypt proper, though I would not be surprised if it were eventually found.

STATUS: There is much ecological variation in the species, but this cannot be classified in subspecific terms.

Members of the genus *Tarucus* are readily recognised by the undersides which are white with a profusion of black spots; along the hindwing edge there are also a series of spots centered with a beautiful metallic green. The male of *T. balkanicus* is blue above, with a number of black spots on the forewing that are missing in the next species. Females are difficult to place correctly, though those of the present species are usually darker than those of *T. rosaceus*. In Egypt this butterfly is, to my mind most surprisingly, limited to the northern Sinai. In the Jordan Valley, a very hot locality, it is often found with the next species on the same *Zizyphus spina-christi* trees (Larsen & Nakamura, 1983), and in Dhofar I have found both with *T. theophrastus* (see section on possible species) on the same *Zizyphus leucodermis* (Larsen, 1983, in which paper I also illustrate the male genitalia of all three). The larval food plants are species of *Zizyphus* and *Paliurus* (Rhamnaceae).

29 *Tarucus rosaceus* Austaut

Lycaena theophrastus var. *rosacea* Austaut, 1885, Naturaliste, 7:141.
Plate 6: 23-25.

DISTRIBUTION: A basically eremic species with a range that stretches from northwestern India, through parts of the Middle East, including Jordan and Israel, to Arabia, Egypt and the Sahel zone of Africa, extending from Mauritania north to the Mediterranean coast.

STATUS: There are no subspecies of this eremic butterfly which in older literature is usually referred to as *T. mediterraneus* Bethune-Baker, a junior synonym, or, erroneously, as the allied *T. theophrastus* Fabricius.

This species (characteristics given under *T. balkanicus* above) is generally fairly common in Egypt wherever its food plant, *Zizyphus,* is found. I have personally taken it in the Siwa oasis, the delta, in Cairo, at Luxor, Aswan and in the Firan oasis in the Sinai. Usually the butterfly is found around the food plants, but it will travel some distance to feed from particularly attractive flowers. Although definitely non-migratory, the species has some powers of dispersal since even isolated *Zizyphus* trees are often colonised. The larvae of the *Tarucus* are easily detected since they eat from only one side of the leaf, leaving the leaf membrane intact. They are usually avidly attended by ants.

30 *Azanus ubaldus* Cramer

Papilio ubaldus Cramer, 1782, Pap. exot., 4:209.
Plate 6: 26-28.

DISTRIBUTION: The range covers most of dry Africa, Arabia and India with very slight penetration of the Palaearctic in North Africa and the eastern Mediterranean. It is somewhat migratory and despite its presence in India is best considered an expansive Afrotropical element.

STATUS: No geographical subspecies can be identified despite the wide range of the species. Many names have been created, including *thebana* Staudinger for Luxor specimens. None should be considered valid.

The male of this tiny species may be recognised from all other Lycaenidae in Egypt by the large, well-defined patch of androconial scales on the upperside of the forewings. The female is normally uniformly brown above. The forewings are more pointed than those of *A. jesous.* The species may be resident in the Aswan area but is probably mainly a migrant in the rest of the country, though occasionally a local population may persist for several years. There are not many records from lower Egypt, and in the Sinai the species is headquartered on the Gulf of Aqaba. In April 1986 I found it very common in an *Acacia* dominated wasteland just north of Aswan together with *Deudorix livia* Klug. The larvae of both feed on *Acacia,* those of *A. ubaldus* often pupating under stones at the foot of the tree.

31 *Azanus jesous* Guérin-Méneville

Polyommatus jesous Guérin-Méneville, 1849, Lef. Voy. Abyss.:383.
Plate 6: 29-31.

DISTRIBUTION: The distribution and the zoogeographical affinities of this species are roughly the same as for the preceding one, though it is not as well established in Egypt.

STATUS: For reasons given by Larsen (1983) the Mediterranean subspecies *gamra* Lederer should not be accepted, nor expected in a migrant species. The clearly defined ssp. *soalalicus* Karsch from Madagascar has been shown to be specifically distinct (Stempffer, 1969). Hence no subspecies should be accepted.

The underside pattern of this butterfly can only be confused with that of

A. ubaldus (above). The male upperside is a delicate violet-blue; the female is brown with lighter coloured discal areas. There appear to be no records of this butterfly from Egypt proper, though it is frequently found on the east coast of the Sinai, especially to the north. It is common in the Jordan Valley and has been found in numbers on the coast of Lebanon (Larsen, 1974). The species is migratory and must turn up in Egypt eventually, not least in the Gebel Elba area, where it is almost certainly resident. The larval food plants are *Acacia* and related trees and scrubs.

32 *Zizeeria karsandra karsandra* Moore

Polyommatus karsandra Moore, 1865, Proc. Zool. Soc. London, 1865:505.
Plate 6: 32-34.

DISTRIBUTION: This is a tiny butterfly with a highly enigmatic distribution pattern, not shared with any other. The range stretches from all of Asia, via eastern Arabia and the Levant to Egypt, then through North Africa to Algeria where it meets the Afrotropical vicariant *Z. knysna* Trimen. Also on Cyprus, Crete and Sicily.

STATUS: The Oriental taxon has often been considered conspecific with the Afrotropical one, but the genitalia and other characteristics differ constantly and most modern authors are agreed to treat them as specifically distinct. A number of subspecies have been described in the Oriental region; their taxonomic validity is doubtful.

Z. karsandra is a tiny butterfly. The male is blue above with an even black margin to the forewings; the female is brown above with a variable degree of blue scaling. The underside is brownish-grey with a profusion of black dots which may tend towards being obsolescent. It is one of the most common butterflies in the cultivated areas of Egypt. One of the chief larval food plants is cultivated alfalfa (*Medicago sative*), but it will feed on many other leguminous plants as well. Benyamini even quotes *Polygonum* and *Tribulus* from the Sinai. It is at its most common in genuine oasis habitats, but it will be found also anywhere with a bit of water or better rainfall than normal for Egypt, not least in the Sinai. All Egyptian material definitely refers to the Oriental species. I made a special point of taking samples all along the Cairo to Aswan road, but as suggested by Wiltshire (1948) it is not improbable that the African species will be found in the Gebel Elba area.

33 *Zizina otis sangra* Moore

Polyommatus sangra Moore, 1866, Proc. Zool. Soc. Lond., 1865:772.
Plate 6: 35-37.

DISTRIBUTION: This tiny species is found on the Pacific Islands through most of the Oriental region, west to about the Region of Karachi. It then recurs, quite unaccountably, in the Siwa oasis of the western desert.

STATUS: Gabriel & Corbet (1949) are emphatic that they were dealing with seven meticulously labelled specimens of the Oriental *Z. otis* and not with the Afrotropical *Z. antanossa*, both on grounds of genitalic configuration and pattern. I have seen the specimens in the British Museum of Natural History and can only agree.

Two males and one female are labelled Siwa, 5.v.35; a further two males and two females are labelled Siwa, 21.vi.35. Males and females of the very similar *Zizeeria karsandra* Moore carry the same labels, so a labelling mistake appears most unlikely. On the underside both sexes lack the black costal spot near the end-cell area that is normally present in *Z. karsandra.* Biogeographically the presence of this species in the Siwa oasis stretches the imagination. Could it have been imported with an Indian strain of animal feedstuff? In this respect it is worth recollecting that a larva of the Oriental Nymphalid *Euthalia aconthea* Cramer was once found on imported mangoes at Suez. That it should be a survivor, solely in Siwa, of an ancient distribution pattern does not seem probable either. However, *Zizeeria karsandra* is also an Oriental species extending to Algeria and its range has a number of disjunctions.

34 *Polyommatus icarus zelleri* Verity

Polyommatus icarus zelleri Verity, 1919, Ent. Rec. J. Var., 31:44.
Plate 7: 3-5.

DISTRIBUTION: This is a common Palaearctic butterfly ranging from North Africa through Europe and the Middle East to Afghanistan.

STATUS: Subspecies are difficult to define. Ssp. *zelleri* is Mediterranean, but the species is in urgent need of a thorough revision.

P. icarus and *Plebejus pylaon* are similar in that the margins of the hindwing underside have a series of well defined orange lunules. However, normally

P. icarus has two black spots at the base of the cell of the forwing underside that are missing in the latter. The males of both species above have the same tone of violet blue; the females of both vary considerably, being brown with a varying degree of blue suffusion and with more or less defined orange marginal lunules. This butterfly is on the very limits of its range in Egypt with a few records from the northern Sinai and a single record from Cairo. The latter was taken near the Gezira Club by Graves; since it has not subsequently been found, one must presume it to have been accidentally imported. Many of its leguminous larval food plants are available in Egypt but presumably summer is too hot to allow the species to colonise permanently.

35 *Agrodiaetus loewii uranicola* Walker

Lampides uranicola Walker, 1870, Entomologist, 5:52.
Plate 7: 9-11.

DISTRIBUTION: There are two broad strains of this species. A desert strain with females that are blackish with white markings (sometimes blue) is found in a number of subspecies from Egypt's eastern desert, through Jordan and Israel to central Arabia, Oman, southwestern Iran and Baluchistan. To the north of this range flies another strain of subspecies whose females are brown with more or less defined red marginal bands, stretching from Turkey through Lebanon, Iraq, Iran to Afghanistan. It seems that the species has differentiated into an eremic and a Palaearctic branch.

STATUS: The Egyptian population has been referred to under a number of names. As shown by Nakamura (1977) the taxon *uranicola* Walker was described from the Sinai and is the oldest name available. The species has been placed in most possible genera by various authors and is probably not well placed in *Agrodiaetus* either. Balletto (in Kudrna, 1986) suggests its placement in *Vacciniina*.

This attractive blue flies in late April and May in a single brood in the eastern desert where it is local, but not uncommon. The large size and the blue colour of the males, wholly without violet, should serve to distinguish it from other similar species. Females may be almost as blue as males, or, more frequently, blackish brown with some lighter markings and no trace of blue. *A. loewii* also flies in April and May at low levels in the Sinai, a month or so later at 2000m. The flight of the male is usually very fast. Most of the females are blackish, but females with varying degrees of intense blue occur. The proportion of blue females varies from colony to colony. The larval food plants are spiny desert *Astragalus*, especially *A. spinosus*.

36 *Plebejus pylaon philbyi* Graves

Plebejus pylaon philbyi Graves, 1925, Trans. ent. Soc. Lond., 73:279.
Plate 7: 6-8.

DISTRIBUTION: The range stretches in a disjunct manner from Spain through Central Europe, then more generally in the Balkans, southern USSR, and the Middle East, south to the Sinai and Riyadh.

STATUS: Ssp. *philbyi* is a small desert subspecies that contrasts markedly with the large, brilliantly coloured ssp. *cleopatra* Hemming from further north. The situation is discussed by Larsen & Nakamura (1983). I would not be surprised if the taxon *pylaon* was eventually found to consist of more than one biospecies.

In Egypt this butterfly is only known from the Sinai, where it exists in scattered colonies except in the coastal zones. It flies to well above 2000 m in the southern mountains. The underside of both sexes is much like that of *Polyommatus icarus*, under which the distinction between the two species is given. There is a single brood about April-May depending on altitude, but at least in Lebanon occasional second brood specimen can be taken (Larsen, 1974). The food plants are desert *Astragalus*, especially *spinosus*, but also others (*fresenii*, *echinus*). The larvae hibernate among the roots of the food plant. In years with environmental stress the larvae are cannibalistic; this allows the population to survive when fresh foliage is in short supply.

37 *Chilades eleusis* Demaison

Lycaena eleusis Demaison, 1888, Bull. Soc. ent. Fr.: lxvi.
Plate 6: 12-13.

DISTRIBUTION: This is an Afrotropical butterfly adapted to the most arid biotopes of the African Sahel. Its distribution is not too well known, but there are records from Ethiopia, Sudan, Chad, Nigeria and Senegal. It extends north to Egypt.

STATUS: After examining the series in the collections of the British Museum of Natural History I find it difficult to accept any subspecies of this butterfly (e. g. *nigeriae*), especially since there is much seasonal variation. Extreme arid forms are tiny. The subspecies *pharaonis* Staudinger, supposedly from Cairo, is definitely a junior synonym of the nominate subspecies.

This species flies in areas of Upper Egypt where the original plant cover remains in unirrigated lands. The bright silvery blue of the male is different from that of the other Egyptian Lycaenidae. It is most common in the Aswan area, flying in a series of broods from March till November. It is apparently most frequent on the islands of the cataract, and the type locality is the Island of Philae, now submerged. It is also known from Abu Simbel. Roell (1959) took it inside the temple compound at Edfu which has now been cleared of vegetation. Despite much searching in Upper Egypt I found it only once, on some waste ground near the temple at Dendera on 18.iv.1987. This is the southernmost authenticated specimen. Staudinger's type for *pharaonis* is labelled Cairo (xii); it is difficult enough to imagine this species occurring in Cairo, but the December date rules out the authenticity since the weather is much too cold for the species to be active at that time of the year.

38 *Freyeria trochylus trochylus* Freyer

Papilio trochylus Freyer, 1844, Neu. Beitr. Schmett., 5:98.
Plate 6: 5-6.

DISTRIBUTION: This tiny butterfly is found practically throughout Africa, in Egypt and the Middle East, Greece and Afghanistan. In India it flies in ssp. *putli* which is sometimes considered specifically distinct. From India it is found to Australia with big distributional gaps that are difficult to interpret.

STATUS: Egyptian material is clearly referable to the nominate subspecies.

This tiny butterfly is relatively common in Egypt though it may be overlooked. It flutters feebly close to the ground. The upperside of both sexes is dark brown with a prominent orange patch at the anal angle of the hindwings. The underside is the usual one of the Polyommatinae, but there is a large orange patch at the anal angle of the hindwing, with black spots crowned by metallic green. The range in Egypt covers the Mediterranean zone, the delta, the Nile Valley south to Aswan, and probably suitable desert localities as well. It is normally found on dry ground and tends to be absent from irrigated areas. In the Sinai it appears to be limited to the mountains of the south, thus being strongly disjunct (Benyamini, 1984). There is a succession of broods most of the year, except in the high Sinai where it flies from April till October. The larval food plants are normally *Heliotropium* and *Indigofera*, but Wiltshire (1948) also suspected *Alhagi*. In the Sinai, Benyamini records *Andrachne* (Euphorbiaceae), a most surprising food plant.

39 ***Pseudophilotes abencerragus coloniarum*** **Turati**

Turania abencerragus coloniarum Turati, 1927, Att Soc. Ital. Sc. Nat., 67:318.
Plate 6: 10.

Pseudophilotes abencerragus nabataeus **Graves**

Turania baton nabataeus Graves, 1925, Trans. ent. Soc. Lond., 1925:62.

DISTRIBUTION: A Palaearctic species of the southern Mediterranean, being found from the Iberian Peninsula through North Africa to Egypt, recurring in the northern Sinai, Israel, Jordan and the Hejaz.

STATUS: The population in the Mediterranean zone west of the Nile is united with the Libyan one as ssp. *coloniarum*. There is then a large distributional gap before reaching the Jordanian/Israeli ssp. *nabataeus*. However, at best these subspecies, as well as ssp. *cavazzi* Turati from Tripolitania, are weakly defined.

The underside of this tiny butterfly is rather like that of *Polyommatus icarus* and *Plebejus pylaon*, but the ground colour is almost pure white and the dark marking pure black. The fringe is clearly chequered. This butterfly seems to be a great rarity in Egypt, the only specimens available being from early this century, all from the Mariut steppe west of Alexandria. Doubtless it occurs elsewhere between Sallum and Alexandria. There are a few records from the northern Sinai as well, the species being scarce also further north. The probable food plants in the Sinai are *Thymus bovei* or *Origanum isthricum*.

40 *Pseudophilotes sinaicus* Nakamura

Pseudophilotes sinaicus Nakamura, 1979, J. Ent.(B), 44:290.
Plate 6: 11.

DISTRIBUTION: This is a Sinai endemic, limited to the southern mountains of the peninsula.

STATUS: This surprising endemic is taxonomically well defined and does not appear particularly closely related to any one of the two other species found in the neighbouring countries.

This butterfly is even smaller than the preceding one. The ground colour, markings and general aspects are much lighter. In the Santa Katarina area it normally flies in a single brood from the beginning of May, with a partial second brood in July. There is a record from Gebel Sirbal, 2000 m, from 17.iii.1976. Normally a species of considerable altitude there are records from the Firan oasis where the normal food plant, *Thymus decussatus*, is not found. The plant grows only above 1800 m. The butterfly is absolutely tiny; small specimens can compete for the title of the world's smallest butterfly. Pupation takes place underground, near the roots of the food plant.

41 *Iolana alfierii* Wiltshire

Iolana alfierii Wiltshire, 1948, Bull. Soc. Fouad 1er Ent., 32:219.
Plate 7: 1-2.

DISTRIBUTION: This species is endemic to the Sinai, the Negev desert and parts of Jordan. From Jordan there is but a single specimen from near Petra (Larsen & Nakamura, 1983). It has recently been refound in Jordan and also collected in the northern Hejaz.

STATUS: Bernardi (1972) considers the species to be a semi-species in the *Iolana iolas* Ochsenheimer super-species. In traditional taxonomic terms it deserves specific status.

This fine endemic species is distributed somewhat patchily throughout the Sinai from 300 to at least 1800 m, but it is not found in the coastal areas. A glance at the plate should show that neither the blue male nor the brown female can be mistaken for any other Egyptian butterfly. Though localised, it is by no means a rare insect. The flight is fast and powerful, often along cliff faces or other prominent topographical features. It is strongly tied to the larval food plant, *Colutea istria*, and is most easily found by looking for the food plants. The larva spends its life inside the puffed up seed pods of the *Colutea*. However, this is also a favourite food plant of *Lampides boeticus*, so breeding larvae from this plant can be disappointing. I saw the butterfly under somewhat unusual circumstances inside the Santa Katarina monastery. On 28.iv.1987 a female was ovipositing on a *Colutea* that was growing on the protective wall of the burning bush, while a male was circling around the tiny courtyard inside the main walls of the monastery.

NYMPHALIDAE

Danainae

42 *Danaus chrysippus chrysippus* Linné

Papilio chrysippus Linné, 1758, Syst. Nat., 10:471.
Plate 3: 6.

DISTRIBUTION: This prominent butterfly has a vast range covering most of Asia and Australia as well as all of tropical Africa. It is migratory and has established outposts in the eastern Mediterranean and may set up temporary populations in the western Mediterranean and even Italy. It recently established a permanent population in the Canary Islands. An almost identical species occurs in Central America.

STATUS: Apart from the populations east of Sumatra it is difficult to determine subspecies of this variable butterfly. The name *aegyptius* Schreber has been applied subspecifically to the populations of Egypt as well as the genotypically variable populations of Africa. The fact that the Egyptian population is almost monomorphic in the nominate form is indicative of Asian rather than African affinities (the form *dorippus* with no black and white on the forewings, and the form *alcippoides* with white hindwings are at frequencies well below 1 per 1000). The population on the Canary Islands is similarly almost monomorphic and cannot possibly be of West African origin, since in that area the species is monomorphic in the form *alcippoides*. For details of the situation in Arabia see Larsen (1983), and for the Mediterranean Larsen (1986); these papers also give references to key literature on this most interesting butterfly.

Confusion between this species and others is only possible with the female of *Hypolimnas misippus*, a deliberate mimic. That species has no markings in the discal area of the hindwings where the male of *D. chrysippus* has four such spots and the female only three. *D. chrysippus* must be the most familiar Egyptian butterfly, being large and conspicuous, as well as being very widespread and relatively common. There is nothing new in this. An anonymous painter in Luxor some 3.500 years ago included it in a hunting scene (depicted in colour by Larsen, 1984a). The species may be found in all parts of Egypt and must be considered a resident despite its migratory potential. Though it cannot breed in the most extreme deserts its migratory habits leads to its turning up even in the most unexpected places. Hayward (1925) and Williams (1929) trace its fluctuations in frequency in Aswan and Giza respectively, but this is not a very fruitful exercise in a species that is at the edge of its climatic range and subject to reinforcement through migration. Wiltshire (1948) listed *Danaus dorippus* as a distinct species from the Gebel Elba, but corrected this the following year. The larvae will feed on practically any species of the family Asclepiadaceae, even a plant as unlikely as *Caralluma*, with *Calotropis procera* as a firm favourite.

Charaxinae

43 *Charaxes hansali hansali* Felder

Charaxes hansali Felder, 1867, Reise Novara, Lep.:446.

DISTRIBUTION: This fine Afrotropical butterfly is found from Kenya to Sudan and the Gebel Elba in Egypt. It also occurs in southwestern Arabia, with an endemic subspecies in Dhofar.

STATUS: In his monograph on the African Charaxinae van Someren (1971) allocated material from Ethiopia and Sudan to the nominate subspecies, but he overlooked the presence of the species in the Asir Mountains and Yemen, as well as the single Egyptian record. There is no doubt that they belong to the nominate subspecies, rather than to ssp. *arabica* Riley from Dhofar.

C. hansali is a large, unmistakable butterfly. The ground colour is blackish brown; there is a prominent cream band across all four wings and the hindwings each have two prominent tails. The female has broader bands than the male, but sexual dimorphism is not marked. This butterfly only qualifies for a place on the Egyptian list because Alfieri collected a large female in the Gebel Elba in the spring of 1928. That it should be found in this area is not surprising. The species is well adapted to arid conditions and is widespread in the Asir Mountains of Saudi Arabia. The larval food plants are normally *Osyris* (Santalaceae), though there are also records from various Salvadoraceae. *Osyris* is not mentioned from the Gebel Elba by Täckholm (1974).

Nymphalinae

44 *Junonia hierta cebrene* Trimen

Junonia cebrene Trimen, 1870, Trans. ent. Soc. London, 1870:353.
Plate 8: 6-7.

DISTRIBUTION: This Palaeotropical species is found throughout the African tropics (except for the rainforest areas), in Arabia and then from the Indian subcontinent to Thailand. The species occasionally reaches the eastern Mediterranean as a migrant.

STATUS: Egyptian and Arabian specimens are of the African subspecies which is well differentiated from the nominate subspecies in Asia.

J. hierta is an unmistakable butterfly with its straw ground colour and with the prominent blue costal spots on the hindwing upperside. The female has more dark markings than the male and the costal spot has a purple tinge (the spot is sometimes obliterated with black scales). This pretty little butterfly is a rare visitor to the temperate zone. There were several records from Lebanon in the early part of this century, but it has not been seen there since. Wiltshire (1948) had only three Egyptian records: Gebel Elba, Kom Ombo and Nag Hamadi so he considered it limited to Upper Egypt. Roell (1959) reports finding several specimens at the temple of Edfu. I found several specimens near Memphis in December, 1974, so it doubtless occurs in Lower Egypt from time to time. According to Benyamini (1984) the species had been quite common in the Santa Katarina area of the Sinai from May to October of 1974, so the two records may be linked. Benyamini also saw it in the Sinai in June of 1976, but it is probably not a resident of any part of Egypt with the exception of the Gebel Elba, though in Upper Egypt temporary populations might survive for several years. The Kom

Ombo record was one of an important migration of the species taking place between 14-28 October, 1938 (Honoré, 1939), and indeed the species is an accomplished migrant (for three such events in Africa see Larsen, 1978a; 1985; and in press). The species is also migratory in India (see Larsen, 1987). The larval food plants are acanthaceous plants, especially *Barleria*, one species of which is found rarely in the eastern desert.

45 *Hypolimnas misippus* Linné

Papilio misippus Linné, 1764, Mus. Lud. Ulr.:264.
Plate 3: 7-8.

DISTRIBUTION: The range covers tropical Africa and Asia, with slight penetration into the temperate zone in Asia and the eastern Mediterranean. The species has established itself in parts of the Caribbean and neighbouring South America; it has been suggested it was accidentally introduced with slaving ships from Africa.

STATUS: There is no geographical variation despite the vast range, probably because the species is strongly migratory.

Among Egyptian butterflies *H. misippus* has the strongest degree of sexual dimorphism; indeed at first sight it is difficult to accept the two sexes as being conspecific. The male is black above, with large egg-shaped white spots, surrounded by a narrow band of shot purple. The female is a near perfect mimic of *Danaus chrysippus*. The species is normally a scarce and fitful migrant into Egypt. Occasionally large populations build up. I found it common on Gezira Island in Cairo in December, 1970. It cannot normally survive a Mediterranean winter, but early this century there was colony near Beirut that existed for several years (Graves, 1911). The female is a wonderful mimic of *Danaus chrysippus*. The common weed, *Portulaca oleracea*, is the preferred larval food plant.

46 *Vanessa cardui cardui* Linné

Papilio cardui Linné, 1758, Syst. Nat., 10:475.
Plate 8: 4.

DISTRIBUTION: This Holarctic migrant is one of the most cosmopolitan of butterflies, having a practically world-wide distribution, being missing only from the tropical rain forests and parts of South America.

STATUS: There is no geographical variation in this species except for the Australian subspecies *kershawi* M'Coy, considered specifically distinct by some authors (Field, 1971). The genus name *Cynthia* is sometimes used, but I see no reason for not considering it congeneric with *Vanessa atalanta*.

A glance at the plate will serve to show that the species cannot be confused with other Egyptian butterflies. Sexual dimorphism is negligible. From time to time *V. cardui* may be found literally anywhere in Egypt, but it is extremely variable in numbers and in the time of its appearance. Often it is quite rare during the hottest months and usually at its most frequent in spring. These changes in frequency are due to extensive migratory activity, but it probably never dies out completely in the Nile Valley. During the period 25.x to 3.xi.1974 I found the species incredibly abundant in Cairo, with hundreds coming to flowers even on the Midan Tahrir. No migratory movement was seen, but on a subsequent visit at X-mas no larvae were found at all, an indication that they moved out of Egypt without breeding (Larsen, 1975). Williams (1929) monitored the seasonal abundance of *V. cardui* at Giza in what must have been some poor seasons. The larva feeds on a bewildering array of plants under normal circumstances, and on practically anything when under stress. *Malva parviflora* is grown in the delta for food (*'khubbeiseh'*) and *Vanessa cardui* has been recorded as a serious pest on this plant (Andres, 1912). I have also found it on cultivated artichokes.

47 *Vanessa atalanta* Linné

Papilio atalanta Linné, 1758, Syst. Nat., 10:478.
Plate 8: 5.

DISTRIBUTION: This Holarctic butterfly is found throughout the western Palaearctic, but it is migratory and the main headquarters would appear to be the Mediterranean basin, from where it flies north in spring to breed in areas where it cannot permanently survive, reaching even north of the Arctic circle on occasion.

STATUS: No subspecies are recognised for this butterfly as is often the case with migrants.

The bright red band on the upperside of all four wings makes *V. atalanta* readily recognisable. Sexual dimorphism is negligible. This butterfly may be fairly common in the northern Delta during the winter months, sometimes being found as far south as Cairo. There is a single record from August, probably an early returnee. In March, 1987 I recorded the species for the first time from the Siwa oasis, one of the most southerly records, though it has sometimes been taken at the Santa Katarina monastery in the Sinai. It does not seem that the species normally breeds in Egypt, though its food plants (mainly *Urtica* and *Parietaria*) are available. I searched extensively for larvae in Alexandria with no result. It thus seems likely that the individuals reaching Egypt are in a state of sexual diapause. However, Andres once found larvae in Alexandria (Andres & Seitz, 1923). Benyamini (1984) expresses surprise that it has not managed to establish breeding populations on *Parietaria alsinifolia* in the Gebel Katarina area.

48 *Melitaea deserticola macromaculata* Belter

Melitaea deserticola macromaculata Belter, 1934, Arb. morph.-tax. Ent., 1:106.
Plate 8: 1-2.

DISTRIBUTION: This is a Palaearctic species with a southern Mediterranean distribution, extending from Morocco and some of the Saharan mountain massifs to Egypt, the Sinai and the Levant. Its northern limit is Lebanon. Ssp. *macromaculata* extends south to the Hejaz, being replaced in the Asir, Yemen and Dhofar by ssp. *scotti* Higgins, and in the Ethiopian highlands by ssp. *abyssinica* Oberthür.

STATUS: The Egyptian populations are best referred to ssp. *macromaculata* though specimens transitional to the nominate subspecies occur. There is a distributional gap as it does not appear to occur in the western desert.

M. deserticola is normally considerably larger than the following species, *M. trivia*, though some females of the latter may be quite large. Both are orange with a profusion of black spots, but *M. deserticola* normally has a full series of black antemarginal spots on the upperside of all four wings; on the hindwings *M. trivia* usually lacks these spots, instead having prominent tear-shaped spots in the discal area (in case of doubt *M. trivia* is characterised by the presence of the lower discocellular vein of the hindwing cell which is missing in *M. deserticola*). All Egyptian records of this butterfly are from east of the Nile in the desert and in the Sinai. Gebel Mokattam used to be a good site, but I have failed to find it in that area. It is widely distributed in the Sinai at levels from 500 to at least 2000 metres; I took it on the last lap of the climb to the summit of Gebel Musa. There are probably usually two broods in spring and early summer, between February and

July. Oddly, the first broods in the Sinai appear to be earlier than the April broods in the eastern desert. *Kickxia aegyptiaca* is probably the most common larval food plant.

49 *Melitaea trivia syriaca* Rebel

Melitaea trivia syriaca Rebel, 1905, Ann. Hofmus. Wien., 21:6.
Plate 8: 3.

DISTRIBUTION: This is a Palaearctic butterfly with a Pontomediterranean/Iranian distribution. It occurs from southeastern Europe throughout the Middle East to Baluchistan.

STATUS: Definition of subspecies in this taxon is difficult since there is much seasonal, ecological and individual variation. Higgins (1941) listed it as ssp. *robertsi*, but in 1956 reversed his view and accepted *syriaca*. It is close to the nominate subspecies, but the coloration is usually lighter. According to Kudrna (1986) the name *trivia* Denis & Schiffermüller is a *nomen nudum* that has been validly replaced with *fascelis* Esper (Paclt, 1949). The International Commission on Zoological Nomenclature has kindly informed me that they consider the name to be valid (A. Gentry, in litt. 15.Oct.1987).

This species is similar to the preceding one under which the distinguishing characters are discussed. In Egypt this butterfly is limited to the Sinai where it is widespread above 700 metres to at least 1500. There is a succession of broods practically throughout the year. The normal larval food plants are species of *Verbascum*; in the South Sinai mountains the impressive, endemic *V. sinaiticum* is the preferred species.

Satyrinae

50 *Pseudotergumia pisidice* Klug

Hipparchia pisidice Klug, 1832, Sym. Phys., pl. 29: figs. 9-10.
Plate 3: 4.

DISTRIBUTION: This is an eastern Mediterranean Palaearctic species, found from southern Turkey to the Sinai. The species has recently been found also in the Hejaz mountains (Pittaway, 1985; Wiltshire, 1986).

STATUS: Contrary to Kudrna (1977) I have no doubt in giving generic status to the group of species that has a most unusual set of appendages to the uncus, the '*uncus antiquus*'. The biogeography of this small genus supports the view. The four species are: *P. wyssi* Christ in the Canary Islands; *P. fidia* Linné in the Iberian Peninsula and North Africa to Tunisia; *P. pisidice* in the eastern Mediterranean; and finally *P. tewfiki* Wiltshire in the high mountains of southwestern Arabia. This is a most unusual, and probably archaic, biogeographical pattern. Benyamini (1984) says that the isolated South Sinai population differs slightly from the ones further north. I have not seen sufficient material, but the naming of the Sinai population would not seem to serve any particular purpose.

The jet black upperside and the silvery underside of this large butterfly are characteristic; males have a patch of androconial scales on the forewing upperside that is lacking in the female. This interesting butterfly has an isolated population in the high mountains of the southern Sinai, but also occurs on the Israel/Sinai border in the Negev. There is a single annual brood, hatching in May. The imago aestivates and re-emerges in October to mate and lay eggs; hence the occasional specimen may be found from May till November. The larval food plants are grasses; Benyamini (1984) suggests that the main food plant is *Piptatherum* but has not found the larvae.

HESPERIIDAE

Pyrginae

51 *Sarangesa phidyle* Walker

Cyclopides phidyle Walker, 1870, Entomologist, 5:56.
Plate 8: 8.

DISTRIBUTION: This is an Afrotropical butterfly found throughout dry tropical Africa and in southwestern Arabia, just extending to Egypt in the Gebel Elba.

STATUS: Often recorded in older literature under the specific names of *eliminata* Holland and *deserticola* Rebel. Both names are best considered synonyms of *phidyle* whose considerable variation is not well expressed in subspecific terms.

S. phidyle is unmistakable in Egypt, especially the ochre underside being unlike that of any other species. It is the only Egyptian butterfly always to settle with the wings held flat. This is a common butterfly, well adapted to arid conditions. In Egypt it is only known from the Gebel Elba area and would not be expected to occur elsewhere in the country since it is a tropical species that cannot survive even moderately severe winters. It has a curious habit of aggregating in caves during the heat of the day. The probable larval food plants are species of Acanthaceae.

52 *Gomalia elma elma* Trimen

Pyrgus elma Trimen, 1862, Trans. ent. Soc. London:288.
Plate 7: 15.

DISTRIBUTION: This is a Palaeotropical species found widely in Africa, Arabia and, more rarely, in peninsular India.

STATUS: The nominate subspecies flies throughout Africa and Arabia, being replaced by ssp. *albofasciata* in India.

The plate will show that the species is distinctive; the white streak on the hindwing upperside is unlike that of any other African skipper. This little butterfly has been found once in the Gebel Elba and is almost certainly limited to that area, though it has a wide range of ecological tolerance. I have found it under oasis conditions in the Musandam Province of Oman, but winters in Egypt proper would probably be too cold. The larval food plants are normally species of *Abutilon*, though other Malvaceae may be used.

53 *Carcharodus alceae alceae* Esper

Papilio alceae Esper, 1870, Eur. Schmett. Abb., I, plate 51, fig. 3.
Plate 7: 17.

DISTRIBUTION: This Palaearctic butterfly is found from Spain through southern and Central Europe to the Middle East and Afghanistan. A disjunct population exists above 2000m in the highlands of the Yemen Arab Republic.

STATUS: Many names have been bestowed on this common skipper, but the European and Levantine populations are best considered as belonging to the nominate subspecies. In Yemen it flies as ssp. *wissmanni* Warnecke in a population that has continuous broods. The North African/South Spain populations have recently been shown to be specifically distinct as *C. tripolinus* Verity (de Jong, 1978).

Normally this species may be distnguished from *C. stauderi* by the more unicolorous pattern and the less developed white markings of the hindwing upperside. In Egypt they only overlap in the Sinai where the colour contrast between forewings and hindwings in the latter tend to be strong. In Egypt this species is limited to the Sinai. There are two populations. One just reaches the northern Sinai from its main Levantine range. The other is disjunct in the southern Sinai mountains, ascending to at least 1800m in the Gebel Musa area. There are a number of broods from March till November. The habitat choice is most often farmland and date gardens where the larvae feed on malvaceous weeds, crops and ornamentals such as *Malva*, *Althaea* and *Alcea*. I would have thought the butterfly could survive in cultivated areas of the delta, but presumably summer temperatures are too high; the Yemen subspecies does not occur much below 2300m.

54 *Carcharodus stauderi ambigua* Verity

Erynnis stauderi race *ambigua* Verity, 1925, Ent. Rec. J. Var., 37:48.
Plate 7: 16.

Carcharodus stauderi ramses Reverdin

Carcharodus ramses Reverdin, 1914, Bull. Soc. Lepid. Genéve, 3:38.

DISTRIBUTION: This is a southern Mediterranean species extending from Morocco to the Levant borders with Iran. There is a distributional gap between Alexandria and the Sinai.

STATUS: The taxonomy of some of the *Carcharodus* is not easy, but most of the literature relevant to Egypt does include this species as *C. stauderi*. There are two distinct and geographically disjunct subspecies in Egypt: ssp. *ambigua* in the central and southern Sinai, linked to the populations of Lebanon, Israel and Jordan; and ssp. *ramses* which is found in the coastal strip west of Alexandria as an endemic subspecies.

The distinguishing characters of this species are covered under the preceding one. According to Benyamini (1984) the species is found in most of the Sinai, excluding the coastal areas, and flies from March till late September in what is probably three broods. Late brood specimens are usually much smaller and much lighter than those of the first brood. Late brood specimens of ssp. *ambigua* thus resemble the first, and apparently only, brood of ssp. *ramses*. The latter appears to be a rather rare subspecies found in the coastal strip, from where there are records from the Mariut steppe and Mersa Matruh. The flight is very fast indeed and the species is rather unobtrusive and so easily missed. The larval food plants in the Sinai have been recorded as *Phlomis aurea* (Labiatae), and at Mersa Matruh as *Phlomis floccosa*.

55 *Spialia doris amenophis* Reverdin

Hesperia amenophis Reverdin, 1914, Bull. Soc. Lép. Genéve, 3:55.
Plate 8: 15-16.

Spialia doris doris Walker

Nisoniades doris Walker, 1870, Entomologist, 5:56.

DISTRIBUTION: This eremic butterfly has a somewhat curious distribution from Baluchistan and southern Iran to most of Arabia and subdesert Jordan. From these it stretches to the Sinai and Egypt, as well as to North Kenya, Ethiopia, Sudan and the Gebel Elba. Finally there is a highly disjunct and subspecifically distinct population in Morocco.

STATUS: The species is quite variable. The Sinai populations are best referred to the nominate subspecies, though they show transitional tendencies towards ssp. *amenophis* in some cases. The Cairo population is clearly distinguished by smaller size and more intensive spotting; it is disjunct from the range of nominate *doris* and fully deserves its subspecific status. I have not seen the Gebel Elba specimens, but they are most likely to belong to the nominate subspecies as do the rest of the East African population.

This species may immediately be recognised by the black ground colour with a profusion of white spots. There are three populations of this butterfly in Egypt. In the northern Sinai there are scattered colonies of the nominate subspecies. They have a succession of broods from April till October. There is another population in the Cairo area (Mokattam Hills,

Helwan, Kassasin and in the neighbouring eastern desert), which has a series of broods from spring till autumn. This is subspecifically distinct (ssp. *amenophis*). Finally, Wiltshire (1948) mentions it from the Gebel Elba area, where it should occur in the nominate subspecies. Most food plant records are of *Convolvulus* (Convolvulaceae); Benyamini records also another member of that family from the Sinai, *Ipomoea stolonifera*.

Hesperiinae

56 *Pelopidas thrax thrax* Hübner

Gegenes thrax Hübner, 1821, Samml. Exot. Schmett., plate 150.
Plate 8: 9-10.

DISTRIBUTION: This is a Palaeotropical butterfly, being found throughout Africa and in most of India. It is well distributed in Arabia, and penetrates the eastern Mediterranean in Egypt and the Levant. Evans (1949) ranges various southeast Asian subspecies as belonging to *P. thrax*, but that is probably doubtful. It might be more realistic to consider the species Afrotropical with some penetration of India.

STATUS: The Egyptian population is clearly referable to the nominate subspecies, rather than to the African ssp. *inconspicua* Bertolini. The species is often quoted under the name *mathias* Fabricius, but there is no chance that this species could ever occur in Egypt. Walker (1870) described *Pelopidas midea* from Cairo, but this is a junior synonym of *thrax* (Evans, 1949).

P. thrax is an extremely powerful butterfly that is characterised by the prominent hyaline spots of the forewings. The male has a prominent white brand that is lacking in the female. This butterfly has some degree of migratory potential without which it would probably not be present in Egypt and the Levant. As it is the species is not at all uncommon in oasis habitats for much of the year, but most commonly in autumn and the late summer months. I have often found it coming to *Lantana* flowers. The larva feeds on various grasses, including cultivated varieties such as rice and sugar cane, but I think it is most unlikely that it will reach serious pest proportions. However, Stephanian (1911) mentions one case where the 'Nili' rice crop on his estate was wholly destroyed by the species.

57 *Borbo borbonica zelleri* Lederer

Hesperia zelleri Lederer, 1855, Verh. zool-bot. Ges. Wien, 5:194.
Plate 8: 11-12.

DISTRIBUTION: This Afrotropical species is found throughout Africa as well as on most of the neighbouring islands. It penetrates the eastern Mediterranean and there are a few records also from the western Mediterranean in Morocco and southern Spain. There is a doubtful record from Arabia (Larsen, 1983).

STATUS: It seems likely that the taxon actually covers more than one biospecies. There are clear genitalic differences between some of the elements included. The taxon *zelleri* was described from 'Syria' by Lederer and should therefore be valid for Egypt whatever the final decision. It is unlikely, though, that the Spanish and Moroccan population actually belong to the same taxon. The name *holli* Oberthür is available for this population.

B. borbonica is similar to *P. thrax* but both sexes may be distinguished by the hindwing underside which has three well defined white spots; *P. thrax* has a variable number of usually ill-defined spots. In addition *B. borbonica* normally has a single spot in the cell of the upper forewing; *P. thrax* always has two such spots. This is not normally a common butterfly. It is known to be migratory and may, actually, not be a permanent resident. Though found in Lebanon, it is certainly not a permanent resident there (Larsen, 1974). According to Graves (1918a) there are two broods; one from May till July, and another in August to September. He certainly gives the impression that the species was then more common than it is now. I found it very common in Alexandria gardens in October, 1972, but have not otherwise seen it in Egypt. The larval food plants are grasses.

58 *Gegenes nostrodamus* Fabricius

Hesperia urbicola nostrodamus Fabricius, 1793:328.
Plate 8: 13-14.

DISTRIBUTION: This is an essentially eremic butterfly found in North Africa, Spain, the Balkans, Turkey, the Middle East, Arabia, and the northwestern parts of the Indian sub-continent. It ascends the Nile, reaching at least the level of Kharthoum, where I have caught it in the Botanical Gardens.

STATUS: Most of the limited variation in this species appears to be ecologically rather than geographically determined and no subspecies should be accepted in the species. Egyptian specimens are generally small and rather light in colour. This should make confusion with the much darker *G. pumilio* difficult if the latter were to turn up in Egypt, as it might well do. Walker (1870) described the female as *Pamphila proclea* from Cairo, but this is simply a junior synonym (Evans, 1949).

This species lacks the hyaline spots that characterise the two previous ones. The male is unmarked brown, the female has some lighter markings that are not hyaline. *G. nostrodamus* is probably the most common of the Egyptian skippers, being found throughout the delta and the Nile Valley, mostly under typical oasis conditions and coming readily to flowers. In the Luxor area it is almost an insect of the shade, and in both 1971 and 1987 I found it common in the gardens of the famous Winter Palace Hotel. It also occurs in the Sinai where Benyamini (1984) found eggs and larvae on *Aeluropus* and *Panicum* grasses.

Species that might possibly occur in Egypt, surprising absentees, or species that have been recorded in error

Papilio machaon Linné is common in Israel, Lebanon and temperate North Africa. There is a population in eastern Arabia, and another in Oman (Larsen, 1983). Given its ability to survive in Arabia, I am surprised that it has not been able to establish itself in the delta and the Nile valley, since there are several potential larval food plants, including *Daucus* and cultivated *Foeniculum*.

Euchloe ausonia melisande Fruhstorfer is common in Israel quite close to the Sinai border and could occur in the northern Sinai. The species is generally allopatric in respect to *Euchloe aegyptiaca* Verity.

Zegris eupheme larseni Pittaway (1985) was first collected near Wadi Rum by me but not described since only a single specimen was involved. Pittaway subsequently found it in the northern Hejaz. It is likely to turn up in a suitable locality in the extreme northeastern parts of the Sinai. Benyamini (pers. comm.) has in fact taken this taxon on the Sinai/Israel border half way between Eilat and el Arish (28.iii.1986) so it should really be included on the Sinai list.

Calopieris eulimene Klug is found from the level of Lake Chad to Kharthoum and the Arabian Tihama coast. I would not be too surprised if it turned up in the extreme South of Egypt. It is often associated with *Colotis ephyia* (see below).

Colotis ephyia Klug may be found in Egypt, since it is found in Sudan and in the Arabian Tihama coast. According to Graves (1915) there are two specimens in the British Museum (Natural History) labelled Upper Egypt, but with no further details. I have been unable to find the specimens.

Colotis evagore Klug is found in Arabia and East Africa, is able to survive in arid areas, and shows some evidence of migration. It may well occur in southern Egypt, and almost certainly in the Gebel Elba.

Colotis eris Klug, *Colotis protomedia* Klug, and *Colotis halimede* Klug were all recorded from Cairo by Walker (1870), but they must be from somewhere in the Red Sea area south of Egypt. All three could well occur in the Gebel Elba, however.

Tarucus theophrastus Fabricius has been quoted many times from Egypt and elsewhere in the eastern Mediterranean. These records are all due to taxonomic and nomenclatural confusion with two other species. Of these *T. rosaceus* Austaut is found in Egypt, *T. balkanicus* Freyer only in the Sinai. The range of *T. theophrastus* covers southwestern Arabia, the Sahel zone of Africa, as well as NW Africa. I would not be surprised if it turned up in the Gebel Elba area.

Chilades galba Lederer is found in the Levant and could conceivably turn up in the northern Sinai.

Pseudophilotes baton Bergsträsser was reported by Wiltshire (1948b) from the Sinai. There is no doubt that this specimen is actually the first record of *P. sinaicus* as stated by Benyamini (1984).

Charaxes jasius Linné was recorded as Egyptian by Staudinger due to a misunderstanding. This record was cited by Andres (1912) and discounted by Graves (1918). I would have thought it extremely unlikely that the species ever occurred in Egypt.

Euthalia aconthea Cramer (as *garuda* Moore) was once found on imported mango at Suez (Storey, 1914). Mango is the common food plant of the species in India. The climate of Egypt is almost certainly too cold for the species.

Junonia orithya Linné is a tropical migrant which I would have expected to reach Egypt on occasion. The Arabian ssp. *here* Lang has been taken in Jordan once, when I caught a specimen in the northern Jordan Valley in autumn of 1983 (Larsen, 1984b). It should reach Egypt from time to time, either in the Arabian subspecies or in the African ssp. *madagascariensis* Guenée.

Melitaea 'didyma' is often mentioned in older literature, but these are all records predating the acceptance of *Melitaea deserticola* as a valid species.

Melitaea phoebe telona Fruhstorfer is able to extend quite far into the arid parts of Israel and Jordan and might well occur in the Sinai.

Acraea doubledayi doubledayi Hewitson is said by Graves (1916) to have been recorded from Egypt; he mentions Aurivillius in this respect. I have not been able to trace any references, but I should not be surprised if it occurred in the Gebel Elba.

Ypthima asterope asterope Klug is a Palaeotropical butterfly with various subspecies in India, the Middle East, Arabia and Africa. It is common in Yemen, Lebanon and Cyprus. It will probably not be found in Egypt, but its absence is surprising since conditions seem right.

Carcharodus lavatherae Reverdin was quoted from the Mariut by Graves (1905) before the description of *Carcharodus stauderi ramses* Reverdin. It is a Central European species that could not possibly occur in Egypt.

Syrichtus mohammed Oberthür is quoted from the Mariut by Andres (1912). This cannot be possible as the species is restricted to Algeria. It is probably a misinterpretation of *Carcharodus stauderi ramses* Reverdin, though the two are not really very similar.

Pelopidas mathias mathias Fabricius is often quoted from the eastern Mediterranean. All records are due to confusion with *Pelopidas thrax* Hübner.

Gegenes pumilio Hoffmannsegg (as *lefebvrei*) was recorded from the Helwan area by Andres (1912) on the apparent authority of Graves. Graves, however, never mentions the species and in his 1918 article criticises Andres for having misinterpreted some of his information. I found a specimen in the Natural History Museum in Paris bearing the label Egypt with no additional details. It cannot be included in the Egypt list at present, though it might well occur in the Sinai.

While the species that might turn up in northern Egypt and the Sinai are quite predictable and the list not very long, the situation in respect of the Gebel Elba area in the extreme southeast is different. Judging from the species already recorded from there it should be possible to find more, possibly many more, but it is not worthwhile making a list of potential species given the anomalous biogeographical position of the area.

Biogeography of the Egyptian Butterflies

It is the intention of the section to analyse the biogeographical composition of the Egyptian butterflies along the lines adopted in my paper on the zoogeographical composition of the Arabian butterflies (Larsen, 1984c), where the definitions of the concepts used are discussed and their history traced. Table 1 below summarises the zoogeographical composition of the Egyptian fauna.

Table 1 Summary of Egyptian butterflies by zoogeographical origin

Zoogeographical Region	No. of species	Percent species
Afrotropical	13	22.5
Oriental	3	5.1
Palaeotropical	8	13.8
Palaearctic	22	37.9
Eremic	11	19.0
Indeterminate	1	1.7
TOTAL	58	100.0

The Afrotropical element

A glance at the Afrotropical section in table 2 below will show that despite being represented with 13 species (nearly a quarter of the total), the Afrotropical element is not well established in Egypt. Seven of the species are migratory and some are not permanent residents anywhere in the country. Five species have managed to bridge the Arabian Gulf and, despite being of obvious Afrotropical origin, are represented also in India, showing them to be exceptionally good colonisers. Finally, three of the species are represented only in the Gebel Elba in the extreme southeastern corner. The only fully resident Afrotropical species in Egypt proper is *Chilades eleusis*.

The Oriental element

Very few Oriental butterflies have managed to penetrate Arabia, Africa and the eastern Mediterranean. The three species in Egypt are quite atypical for Oriental species. *Zizeeria karsandra* stretches from the Oriental Region through Baluchistan, eastern Arabia, and the Levant to Egypt and Algeria, a unique distribution pattern. *Zizina otis* is limited to the Siwa oasis and the nearest colony is in Pakistan. *Madais fausta* appears to be of Indian origin, but is highly migratory in parts of Arabia and the Middle East.

The Palaeotropical element

A large proportion of Palaeotropical butterflies are also migratory (see discussion in Larsen, 1984c), and this is true for seven of the eight Egyptian representatives which are all widely distributed species which also penetrate the Levant on occasion. The eighth species is an arid zone skipper found only in the Gebel Elba.

The Palaearctic element

With 22 species the Palaearctic element is by far the largest, comprising almost forty percent of the total. Again, a glance at table 2 will show that it is also an element that is not well established. Seven of the species are migratory, though a few of these are well established. No less than eight have only been recorded from the Sinai, two of which are endemic, and several of which barely penetrate the peninsula. Others are the weak easternmost irradiation of the North African/Iberian speciation centre found only in the coastal strip up to Alexandria. Only *Euchloe belemia* and *Melitaea deserticola* are more widely established as well as non-migratory. *Artogeia rapae*, the most common of the Palaearctic species, is migratory and basically synanthropic.

The eremic element

Eleven of the Egyptian species may be classed as eremic under the definition discussed in Larsen (1984c). These are butterflies centered on the desert zone stretching from Mauritania to the northwestern corner of the

Indian subcontinent and in some cases to Central Asia. By and large these are the only species to survive permanently in the deserts. Some of them have not been found in the western desert, but I suspect that most will eventually be found even there. These species have finely tuned diapause mechanisms or feed on plants with very deep root systems that grow leaves even in years with no rain. The two species of *Apharitis* can apparently manage by eating the early stages of ants inside ants' nests. Compared to the Arabian eremic fauna there are few absentees (Larsen, 1984c: table 13).

Table 2 below classifies the butterflies known from Egypt according to their zoogeographical origins along with some brief notes on any anomalous characteristics.

TABLE 2 Egyptian butterflies by zoogeographical origin

Zoogeographical Region	Notes		
Afrotropical species			
Colotis phisadia	I	m	
Colotis chrysonome			GE
Colotis danae	I		
Colotis ?liagore			
Deudorix livia		m	
Anthene amarah	I	m	
Syntarucus pirithous		m	
Azanus ubaldus	I	m	
Azanus jesous	I	m	
Chilades eleusis			
Charaxes hansali			GE
Sarangesa phidyle			GE
Borbo borbonica		m	
Oriental species			
Madais fausta		m	
Zizeeria karsandra			
Zizina otis 4)			
Palaeotropical			
Anaphaeis aurota		m	
Catopsilia florella		m	
Lampides boeticus		m	
Danaus chrysippus		m	
Junonia hierta 3)		m	
Hypolimnas misippus		m	
Gomalia elma 3)			GE
Pelopidas thrax		m	

Palaearctic species		
Papilio saharae 1)		
Artogeia rapae	m	
Pieris brassicae	m	
Pontia daplidice	m	
Euchloe belemia		
Colias croceus	m	
Satyrium jebelia		S
Tomares ballus		
Lycaena phlaeas	m?	
Lycaena thersamon		S
Polyommatus icarus		
Plebejus pylaon		S
Pseudophilotes abencerragus		
Pseudophilotes sinaicus		S
Iolana alfierii		S
Vanessa atalanta	m	
Vanessa cardui	m	
Melitaea deserticola		
Melitaea trivia		S
Pseudotergumia pisidice		S
Carcharodus alceae		S
Carcharodus stauderi		
Eremic species		
Pontia glauconome		
Euchloe aegyptiaca		
Euchloe falloui		S?
Elphinstonia charlonia		
Apharitis acamas		
Apharitis myrmecophila		
Tarucus balkanicus		S
Tarucus rosaceus		
Agrodiaetus loewii		
Spialia doris		
Gegenes nostrodamus		
Anomalous species		
Freyeria trochylus 2)		

Symbols and footnotes: I = Afrotropical species also penetrating to India; m = migratory; GE = in Egypt limited to Gebel Elba; S = in Egypt limited to the Sinai; 1) the species could be considered an eremic representative of a Palaearctic group; 2) found in most of Africa and Asia as well as the Balkans and the Middle East; 3) the Afrotropical subspecies in Egypt; 4) a population in Siwa oasis only, then from Karachi eastwards.

Endemicity

Under the circumstances it is hardly surprising that the level of specific or subspecific endemicity in Egypt should be very low, in fact limited to the following five cases:
Satyrium jebelia - endemic to the Sinai
Tomares ballus mareoticus - ssp. endemic to the Mariut
Pseudophilotes sinaicus - endemic to the Sinai
Spialia doris amenophis - ssp. endemic to eastern desert
Pseudotergumia pisidice - slightly differentiated in the Sinai.

The Distribution of Butterflies within Egypt

The geographical positioning and ecological conditions of Egypt are such that very few butterflies can be found in all parts of the country. As shown in table 4 below, only *Deudorix livia*, *Lampides boeticus*, *Syntarucus pirithous*, *Tarucus rosaceus* and *Danaus chrysippus* share this honour. At the other extreme no less than ten species have only been recorded in the Sinai, four only in the Gebel Elba, and two only from the coastal strip. This is illustrative of Egypt's positioning as a meeting point for the hardier species of several different faunal regions.

Table 3 below outlines the number of butterfly species that are found in each of the seven ecological zones discussed in the introduction to this review. It is not easy to make a satisfactory classification for each of the butterflies as to its status in a given zone. As in table 4 below the following criteria have been applied:

Resident breeding species are those that maintain permanent breeding populations in the zone; though migration may also be involved the species is always present.

Uncertain breeding status is used for species which are not normally migratory and which have been recorded from the zone, but where it is uncertain whether there are permanent breeding populations.

Certain to occur is used for species that have not been recorded from a zone, but where distribution and ecological requirements indicate that it has been overlooked. The category is mainly applied to the ill-explored western deserts and to the Gebel Elba. Certain should obviously always be understood as almost certain.

Occurs as *migrants* only is used to denote a species that would not be present except for migration; though they may survive for years, they do not establish permanent populations.

While the Egyptian fauna is well known in respect of the species present, more data on permanent breeding status and migration is needed before the figures given in the table can be considered anything but a provisional review. It is nonetheless felt worthwhile to give the figures and to make some comments thereon.

Table 3 shows that there is not much difference in the total number of recorded butterflies from zone to zone, with the exception of the Sinai. However, while the figures for six of the zones cannot be far off the true numbers, that of the Gebel Elba is almost certainly an underestimate. Only the species already known from Egypt are included in the category 'certain to occur', but there are definitely several, perhaps even many, that have not previously been recorded and that will not be found outside of Gebel Elba. However, the similar total figures camouflage a rather different faunal composition in each of the zones. These differences will be summarised below.

Table 3 The number of butterflies within the seven main ecological zones in Egypt.

Status of species	Coastal strip 1	Western desert 2	Lower Nile 3	Upper Nile 4	Eastern desert 5	Gebel Elba 6	Sinai 7
Resident breeding species	12	7	13	15	14	11	34
Uncertain breeding status	4	4	2	1	6	0	1
Certain to occur	0	5	0	0	2	17	0
Occurs as migrants only	6	6	8	5	7	0	9
TOTAL	22	22	23	21	29	28	44

Source: Summary of table 4 below.
Note: For definition of status categories see paragraph preceding the table. Benyamini (pers. comm.) believes that three species that I consider migrants are permanent in the Sinai.

The following paragraphs will briefly outline the fauna of each of the seven ecological zones, highlighting their peculiarities in relation to the neighbouring zones:

In the coastal strip 21 species have been recorded, a few only as migrants, constituting 36 % of the total fauna. *Pieris brassicae* and *Tomares ballus* are unique to this environment, and two further species are shared only with the Sinai. These are all Palaearctic species. The remainder consists of the hardier tropical species and migrants. I may have underestimated the number of eremic species that could occur, but the considerable amount of material available from the Mariut indicates that the true desert species cannot cope with the coastal climate.

The western desert has 22 species, though five of these are in the category of 'certain to occur'. The reason for this relatively high figure is that some species survive in oasis environments and would otherwise be absent except as rare strays. Most of the true eremic species are present or certain to occur. The only real surprise on the list is that of *Zizina otis* from the Siwa oasis. One species, *Chilades eleusis*, is limited to the southern part of the desert.

The oasis biotope of the lower Nile valley contains 23 species, mainly the Palaearctic species and the tropical migrants. *Euchloe belemia*, *Lycaena phlaeas*, *Polyommatus icarus* and *Vanessa atalanta* do not ascend to Upper Egypt. The second and third of these butterflies probably do not have resident status, and nor, of course, does *Vanessa atalanta*. Nearly all the true eremic butterflies are conspicuously absent from the well irrigated lower Nile and delta.

The upper Nile valley has only 21 species. It differs from the lower Nile in faunal composition through losing the four Palaearctic species mentioned above. It gains *Pontia glauconome* in the uncultivated parts of the valley, as well as *Chilades eleusis*. Thus 19 of the species overlap between the lower and upper Nile. However, a few of the tropical species appear to be permanent residents of the upper Nile, but not of the lower.

The eastern desert with 29 species, 50 % of the total, scores relatively high. The reason lies in the higher and more predictable rainfall compared to the western desert, giving rise to vegetation able to support more species. The only species of the western desert not found in the eastern are *Elphinstonia charlonia* and *Zizina otis*. In contrast, *Colotis phisadia*, *Madais fausta*, *Agrodiaetus loewii*, *Melitaea deserticola* and *Spialia doris* appear to be missing

Table 4 The distribution of butterflies within the seven main ecological zones in Egypt.

FAMILY Species	Coastal strip 1	Western desert 2	Lower Nile 3	Upper Nile 4	Eastern desert 5	Gebel Elba 6	Sinai 7
PAPILIONIDAE							
Papilio saharae	o	•			•		•
PIERIDAE							
Artogeia rapae	•	o	•	•	o		•
Pieris brassicae	m						
Pontia daplidice	m	m	m	m	m		•
Pontia glauconome	•	•		o	•	?	•
Euchloe aegyptiaca	•				•		•
Euchloe falloui		?			?		•
Euchloe belemia	•		•		o		
Elphinstonia charlonia	•	•					•
Colotis phisadia					o	?	•
Colotis chrysonome						•	
Colotis danae						•	
Colotis ? liagore						•	
Madais fausta					•	?	•
Anaphaeis aurota		m	m	m	m	?	m
Catopsilia florella		?	m	m	m	?	m
Colias croceus	•	o	•	•		o	•
LYCAENIDAE							
Deudorix livia	m	?	m	•	m	•	m
Satyrium jebelia							•
Tomares ballus	•						
Apharitis acamas		?			•	?	•
Apharitis myrmecophila		?			•	?	•
Lycaena phlaeas	o		o				•
Lycaena thersamon							•
Anthene amarah						•	m
Lampides boeticus	•	m	•	•	m	?	•
Syntarucus pirithous	o	o	•	•	•	?	•
Tarucus balkanicus							•
Tarucus rosaceus	o	•	•	•	•	?	•
Azanus ubaldus			m	•	m	?	m
Zizeeria karsandra	•	•	•	•	•		•
Zizina otis		•					
Polyommatus icarus			o				o
Agrodiaetus loewii					•		•
Plebejus pylaon							•
Chilades eleusis		•		•	?		

Freyeria trochylus	•	?	•	•	•	?	•
Pseudophilotes abencerragus	•						•
Pseudophilotes sinaicus							•
Iolana alfierii							•
NYMPHALIDAE							
Danaus chrysippus	m	m	•	•	•	•	•
Charaxes hansali						•	
Junonia hierta			m	m		•	m
Hypolimnas misippus			m	m		?	
Vanessa cardui	m	m	•	•	m	?	m
Vanessa atalanta	m	m	m				m
Melitaea deserticola					•		•
Melitaea trivia							•
Pseudotergumia pisidice							•
HESPERIIDAE							
Sarangesa phidyle						•	
Gomalia elma						•	
Carcharodus alceae							•
Carcharodus stauderi	•						•
Spialia doris					•	•	•
Pelopidas thrax			•	•	o	?	
Borbo borbonica			•	?		?	
Gegenes nostrodamus			•	•	o		•

Notes: The following symbols are used:
- • resident species in the zone in question
- o recorded from the zone but resident status uncertain
- ? not recorded but almost 100% certain to occur
- m present only as a migrant

For more detail see the paragraph preceding table 3.
Benyamini (pers. comm.) believes that three species that I consider migrants are permanent in the Sinai.

from the western desert. Finally, ecological conditions permit breeding of some tropical or Palaearctic species, at least in spring.

The special nature of the Gebel Elba range will already be apparent. Six of the eleven species known for sure to exist there have been found nowhere else in Egypt. My list of 17 species 'certain to occur' is conservative, but in addition there will be species not yet recorded from Egypt. I would not be surprised if the tentative figure of 28 were to be doubled. Biogeographically the Gebel Elba area has little in common with the rest of Egypt.

That the Sinai peninsula should have as many as 44 species of butterflies probably comes as a surprise to many. It is by far the richest Egyptian zone. The reasons are threefold: 1) the basic eremic habitat contains nearly all species found in the deserts of Egypt proper; 2) the high mountains of the southern Sinai and parts of the northern Sinai support many of the Egyptian Palaearctic species, plus a number that are not found in Egypt proper, including two endemics; 3) the proximity of the Hejaz mountains to the eastern coast of the Sinai allows the penetration of tropical species. The fourteen Egyptian species missing from the Sinai include six or seven from the Gebel Elba, two or three of the Mariut Palaearctic species, one or two migrants, *Zizina otis* and *Chilades eleusis*.

Migrant Butterflies in Egypt

Migration plays a major role in the butterfly faunae of the arid zone where rainfall is low, but above all unpredictable. Out of Egypt's 58 species of butterfly, no less than 22 (38%) are migratory. If the eight Palaearctic species limited to the Sinai and the four tropical species limited to the Gebel Elba are removed, this percentage rises to nearly fifty.

In the following paragraph I have attempted to divide the Egyptian migrant butterflies into three categories, based on my three earlier papers on butterfly migration in the Middle East and Arabia (Larsen, 1975; 1976; 1982c). Such a classification is difficult and to some extent imprecise since species may show different migratory behaviour in different localities. Thus, *Madais fausta* is strongly migratory in the Levant, building up large populations over several broods every autumn, only to die out every winter. Yet in India this species, in two distinct subspecies, is extremely sedentary. Some migrants may build up populations that survive for years in areas where they cannot permanently exist. *Danaus chrysippus* in recent years has established flourishing colonies in southern Spain and is being hailed as a new resident species; I am quite convinced that one bad winter will eventually kill off these colonies as happened once to a breeding colony in Italy (Larsen, 1986). Similarly, a colony of *Hypolimnas misippus* survived for some years in Beirut earlier this century.

Regular migrants

This category represents species which regularly undertake long-distance directional movements, and where migration is an essential component in their life histories or survival strategies, though not necessarily everywhere within the range of the species. Most of the members of this category would probably become extinct in Egypt in the absence of active migratory movements.

Pieris brassicae
Pontia daplidice
Madais fausta
Anaphaeis aurota
Catopsilia florella
Colias croceus
Lampides boeticus
Danaus chrysippus
Vanessa cardui
Vanessa atalanta

Moderate migrants

This category represents species which have frequently been noted as long-distance directional migrants, but where long-distance migration is not an essential part of their life cycle and survival strategies. Though migratory, most have permanent resident populations in Egypt.

Artogeia rapae
Colotis phisadia
Deudorix livia
Syntarucus pirithous
Azanus jesous
Azanus ubaldus
Junonia hierta
Hypolimnas misippus
Pelopidas thrax

Weak migrants

This category represents species which have not been observed in typical migrations, but whose distribution patterns, lack of geographical differentiation, unpredictability, and ability to colonise to the very limits of their potential habitat indicate that they are migratory or strongly colonising.

Plate 1

1. Plantation in the Delta near Alexandria.
2. Tropical vegetation on rough ground on the Nile near Luxor.
3. General view the Western Desert - many areas have much less vegetation and are wholly devoid of butterflies.

1
2
3

Plate 2

1. Vegetation in one of the small wadis of the Eastern Desert.
2. The Firan Oasis is the largest of its kind in Sinai (700 m +).

Plate 3

1. *Papilio saharae saharae* male UPS (no. 1)
2. *Catopsilia florella* male UPS (no. 16)
3. - - female UPS
4. *Pseudotergumia pisidice* female UPS (no. 50)
5. *Pieris brassicae brassicae* female UPS (no. 3)
6. *Danaus chrysippus chrysippus* male UPS (no. 42)
7. *Hypolimnas misippus* male UPS (no. 45)
8. - - female UPS

1

2

3

4

5

6

7

8

Plate 4

1. *Artogeia rapae leucosoma* male UPS (no. 2)
2. - - female UPS
3. *Euchloe aegyptiaca* male UNS (no. 6)
4. *Pontia daplidice daplidice* male UPS (no. 4)
5. - - female UPS
6. - - female UNS
7. *Pontia glauconome* male UPS (no. 5)
8. - - female UPS
9. - - female UNS
10. *Euchloe belemia* male UPS (no. 7)
11. - - female UPS
12. - - female UNS
13. *Euchloe falloui* male UPS (no. 8)
14. - - female UPS
15. - - female UNS
16. *Anaphaeis aurota* male UPS (no. 15)
17. - - female UNS

1
4
7
2
5
8
3
6
9
10
13
16
11
14
12
15
17

Plate 5

1. *Colotis chrysonome* male UPS (no. 11)
2. - - female UPS
3. - - male UNS
4. *Colotis phisadia phisadia* male UPS (no. 10)
5. *Elphinstonia charlonia charlonia* male UPS (no. 9)
6. - - male UNS
7. *Colias croceus* male UPS (no. 17)
8. - - female UPS (white form)
9. *Madais fausta fausta* male UPS (no. 14)
10. *Colotis danae eupompe* male UPS (no. 12)
11. *Colotis liagore* male UPS (no. 13)
12. - - female UPS
13. *Deudorix livia* male UPS (no. 18)
14. - - female UPS
15. - - male UNS
16. *Lycaena thersamon* male UPS (no. 24)
17. - - female UPS
18. *Lycaena phlaeas* male UPS (no. 23)

1
2
3
4
5
6
7
8
9
10
11
12
13
14
15
16
17
18

Plate 6

1. *Apharitis acamas* female UPS (no. 21)
2. - - male UNS
3. *Apharitis myrmecophila* male UPS (no. 22)
4. *Satyrium jebelia* male UPS (no. 19)
5. *Freyeria trochylus* female UPS (no. 38)
6. - - female UNS
7. *Tomares ballus mareoticus* male UPS (no. 20)
8. - - - female UPS
9. - - - male UNS
10. *Pseudophilotes abencerragus* male UPS (no. 39)
11. *Pseudophilotes sinaicus* male UPS (no. 40)
12. *Chilades eleusis* male UPS (no. 37)
13. - - female UPS
14. *Anthene amarah* male UPS (no. 25)
15. - - female UPS
16. - - male UNS
17. *Syntarucus pirithous* male UPS (no. 27)
18. - - female UPS
19. - - male UNS
20. *Tarucus balkanicus* male UPS (no. 28)
21. - - female UPS
22. - - male UNS
23. *Tarucus rosaceus* male UPS (no. 29)
24. - - female UPS
25. - - male UNS
26. *Azanus ubaldus* male UPS (no. 30)
27. - - female UPS
28. - - male UNS
29. *Azanus jesous* male UPS (no. 31)
30. - - female UPS
31. - - female UNS
32. *Zizeeria karsandra* male UPS (no. 32)
33. - - female UPS
34. - - male UNS
35. *Zizina otis* male UPS (no. 33)
36. - - female UPS
37. - - male UNS

1
2
3
4
5
6
7
8
9
10
11
12
13
14
15
16
17
18
19
20
21
22
23
24
25
26
27
28
29
30
31
32
33
34
35
36
37

Plate 7

1. *Iolana alfierii* male UPS (no. 41)
2. - - female UPS
3. *Polyommatus icarus* male UPS (no. 34)
4. - - female UPS
5. - - male UNS
6. *Plebejus pylaon* male UPS (no. 36)
7. - - female UPS
8. - - male UNS
9. *Agrodiaetus loewii* male UPS (no. 35)
10. - - female UPS
11. - - male UNS
12. *Lampides boeticus* male UPS (no. 26)
13. - - female UPS
14. - - male UNS
15. *Gomalia elma elma* male UPS (no. 52)
16. *Carcharodus stauderi ambigua* male UPS (no. 54)
17. *Carcharodus alceae alceae* male UPS (no. 53)

1
2
3
4
5
6
7
8
9
10
11
12
13
14
15
16
17

Plate 8

1. *Melitaea deserticola macromaculata* female UPS (no. 48)
2. - - - male UNS
3. *Melitaea trivia syriaca* male UPS (no. 49)
4. *Vanessa cardui* male UPS (no. 46)
5. *Vanessa atalanta* female UPS (no. 47)
6. *Junonia hierta cebrene* male UPS (no. 44)
7. - - - female UPS
8. *Sarangesa phidyle* male UPS (no. 51)
9. *Pelopidas thrax thrax* male UPS (no. 56)
10. - - - female UNS
11. *Borbo borbonica zelleri* male UPS (no. 57)
12. - - - female UPS
13. *Gegenes nostrodamus* male UPS (no. 58)
14. - - female UPS
15. *Spialia doris amenophis* male UPS (no. 55)
16. - - - female UPS

1
4
2
5
3
6
9
11
10
12
7
13
14
8
15
16

Colotis chrysonome
Lycaena phlaeas
Anthene amarah

Butterfly migration, expressed by many individuals of one or more species flying in the same direction for several days under the directional control of the insects involved, covers several different underlying phenomena.

a) Some species migrate between breeding sites and diapause sites. This occurs in some Lebanese butterflies (Larsen, 1975), and seems to have a parallel in California, where the ecological and topographical features are somewhat similar (Shapiro, 1980). These migrations are not very long, but they bring the butterflies from and to areas with very different ecological conditions. It is characteristic for this type of migration that the same specimen makes the journey in both directions. The incredible migrations of the American Monarch (*Danaus plexippus* Linné), where the same individual flies from eastern USA or Canada to a few roosting sites in Mexico and back again in spring after hibernation can be interpreted as an extreme example of this type of migration. None of this type of migrant is known from Egypt.

b) A number of species regularly migrate in response to some sort of seasonal stimulus. Thus, in Egypt *Vanessa atalanta* is only a winter visitor, though possibly occasionally breeding. Every spring it flies north to breed in Europe in a succession of broods, each one extending further north, sometimes reaching the Arctic Circle. In autumn some specimens return to Egypt for hibernation, but they are not the same individuals that left in spring. Only *V. atalanta* in Egypt belongs to this group, though *Colias croceus* and *Pieris brassicae* show some similarity.

c) Most migrant species, especially those that move in large numbers, seem to adopt a strategy of opportunistic nomadism. Migration ensures that specimens will turn up and take advantage of good temporary breeding opportunities where-ever they occur. Such a strategy is of particular relevance in arid zones where rainfall and breeding possibilities are unpredictable. George Popov (pers. comm.) saw vast numbers of the larvae of *Vanessa cardui* in the desert of Saudi Arabia, near Medain Saleh in April, 1947. They were competing with locusts for food and actively advanced through the exceptionally rich vegetation of annuals with the young locust hoppers. It was a year of exceptional rainfall in the desert but no food would have been available on which the resulting adults could have laid eggs. I have no doubt that a migration ensued, but the interesting thing is that no specific flight direction for the migration would necessarily be the best: west - breeding in the Nile Valley would be possible; north - the plains of Anatolia might be perfect for breeding; south - in a year of good rains the Yemen mountains or Ethiopia would be suitable; east - Oman, the Zagros or Baluchistan might be ideal.

d) Finally the weak migrants, and probably also some from the other categories that find ecological conditions in Egypt suitable, probably only have the migratory response triggered by special circumstances.

My conclusion is that most migration in Egypt is a dispersive survival strategy in a difficult and unpredictable environment, evolved instead of the alternative strategy which is a high degree of specialisation to arid conditions. This is perhaps best illustrated through the two closely related Egyptian butterflies, *Pontia daplidice* and *Pontia glauconome*. The first is a typical opportunistic migrant, colonising and breeding in oases in both Egypt and Arabia. The second is sedentary, having specialised in desert life by, among other mechanisms, being able to diapause in the pupal stage for several years.

Butterflies as Pests

Most butterflies do not normally attain serious pest status, but a remarkably high proportion of Egyptian butterflies do feed on plants that are beneficial to human beings. In part this is a reflection of the fact that nature in Egypt has been so radically transformed by human activity during a unbroken period of more than 5000 years. The following species have been recorded as actual or potential pests.

Papilio saharae can feed on fennel and in extreme cases do damage to this insignificant crop.

Pieris brassicae can do serious damage to all types of cabbages but it is only a rare stray in Egypt.

Artogeia rapae is a serious pest on cabbages as it is often very common and has good powers of dispersal.

Catopsilia florella feeds on all types of *Cassia* and always has the potential for reaching pest proportions.

Colias croceus in Egypt feeds mainly on alfalfa (*Medicago sative*). It may be very common but the plant is so profuse in its growth that serious damage will be rare.

Deudorix livia is a highly polyphagous insect that often attacks tree crops such as loquat and dates; it would almost certainly also accept peaches and apricots. It bores inside the fruits and their kernels. Normally the species is not all that common, but occasional outbreaks of pest proportions could well occur.

Lampides boeticus feeds on many genera of the Leguminosae. In Egypt it is particularly fond of cultivated peas and beans on which it is often a serious pest, reducing both the quantity and quality of the crop.

Zizeeria karsandra and *Zizina otis* both feed chiefly on alfalfa (*Medicago sative*) but it is doubtful whether they can do serious damage to this vigorous crop.

Hypolimnas misippus often feeds on *Portulaca oleracea*, used for salads in many parts of the world.

Vanessa cardui is highly polyphagous and may occur in countless numbers. Serious damage has been reported on *Malva parviflora* which is used as a vegetable. Artichokes may also be attacked. Cotton has been known to be defoliated when the density of larvae becomes particularly high.

Pelopidas thrax feeds on most types of grasses. There is at least one record of it doing serious damage to rice (*Oryza*) (Stephanian, 1911) in Egypt. Moderate damage to sugar cane has also been reported elsewhere.

Borbo borbonica is a potential pest on the same crops as those of the preceding species, but *P. thrax* is normally much more common.

Concluding Remarks

The picture that emerges of the Egyptian butterfly fauna is not surprising. The severe climate with little, and unpredictable, rainfall ensures that the total number of species stands at a modest 58. The geographical positioning of Egypt ensures that many of these species are only found at the very frontiers of the country.

About ten of the Palaearctic Egyptian species are restricted to the Sinai Peninsula, a further two are restricted to the Mediterranean coastal strip west of Alexandria. Two more are shared between the coastal strip and the Sinai. Six Afrotropical species are found only in the Gebel Elba in the monsoon zone of the extreme southeast of the country; a few of these are shared with the eastern coast of the Sinai. Two or three more have been recorded on only one or two occasions. To this may be added a few species that almost certainly could not survive permanently in Egypt without migration. Thus, about half the total fauna is not really integral to Egypt as a whole.

The remaining core of butterflies is weighted in favour of the eremic desert specialists and the most hardy and successful of the adventive and migratory species of the tropics and, to a lesser degree, the Palaearctic.

There is no evidence that the Nile has recently acted as a means of communication between the Mediterranean and the Afrotropical Region, and I suspect that it never has. The well documented evidence that many Afrotropical mammals inhabited Egypt up to very recent times was probably due to a general northwards extension of the African savanna during more mesic times, and not to the use of the Nile as a narrow corridor.

Even a generous (leaving out the Gebel Elba) list of additional species that might be expected in Egypt is a short one. Further studies to assess which species are present are still desirable, not least in the western desert. However, the most interesting prospects for further studies into Egyptian butterflies would be an assessment of the extent to which the various species are truly resident in the various zones. This should be coupled with an attempt to elucidate the critical factors that are responsible for the absence or presence of a given species.

This review forms yet a contribution to the strong increase in our knowledge of East Mediterranean and Arabian butterflies that has taken place during the last fifteen years. The faunae of Lebanon, Jordan, Israel, the Sinai, and the whole of Arabia, and Cyprus have been subjected to intensive and up-to-date treatment. A review of the faunae of Libya and Sudan would be very welcome, but given the present political circumstances, this will probably take some time. That an authoritative treatment of the Algerian fauna is not available is somewhat surprising; it would appear to be a most attractive prospect for a French speaking entomologist.

Acknowledgements

This paper would not have been possible without the support of the Danish Carlsberg Foundation which financed both some of my time and a six week exploratory trip to Egypt. The Foundation also sponsored my trip to Jordan in 1977 which was the springboard for my subsequent researches into Arabian butterflies.

I am indebted to the British Museum (Natural History) for access to their collections, and to Richard Vane-Wright and Philip Ackery for assistance and support. The facilities of the library of the Royal Entomological Society of London and the help of its librarian, Brenda Leonard, were as always invaluable.

Tony Pittaway kindly read the entire manuscript and as usual provided additional data and helpful suggestions, as did Duby Benyamini, who was kind enough also to supply unpublished data.

Mrs. A. Gentry of the International Commission for Zoological Nomenclature kindly advised on the status of the name *Papilio trivia* Denis & Schiffermüller.

My thanks are also due to Bettina and Leonardo Leopoldi for their help and company in the Siwa Oasis. Mohammed Heidar kindly conducted me on a tour of the northern part of the oasis. Nancy Fee deserves credit for keeping me awake during long desert traverses, and much more.

The plates were photographed by the Photographic Section of the British Museum (Natural History) to their customary high standards.

Peder Skou of Apollo Books and Leif Lyneborg greatly facilitated the editorial process for which I am most grateful.

References

ALFIERI, A. 1942. Quatre Lépidoptères nouveaux pour la faune du Sinai. Bull. Soc. Fouad 1er Ent., 26:163-164.

ANDRES, A. 1912. Verzeichniss der bis jetzt in Ägypten beobachteten Schmetterlinge. Bull. Soc. Ent. Egypte, 3:53-114.

ANDRES, A. 1930. A new form of Chrysophanus phlaeas L. from Egypt. Bull Soc. Ent. Egypte, 14:212-213.

ANDRES, A. & SEITZ, A. 1923. Die Lepidopteren-Fauna Ägyptens. Senckenbergiana, 5:1-54.

ANDRES, A. & SEITZ, A. 1924. Ibid, 6:1-83.

ANDRES, A. & SEITZ, A. 1925. Ibid, 7:54-61.

AWADALLAH, A. M., AZAB, A. K., & NAHAL, A. K. N. 1970. Studies on the pomegranate butterfly, Virachola livia (Klug). Bull. Soc. Ent. Egypte, 54:545-567.

BACK, W. 1982. Über die Biologie von Euchloe charlonia amseli Gross & Ebert 1975 (Lep.; Pieridae). Atalanta, 13:48-60.

BACK, W. & EITSCHBERGER, U. 1976. Die Unterarten von Euchloe charlonia (Donzel, 1842). Atalanta, 7:142-174.

BENYAMINI, D. 1983. Distribution list of the butterflies of Israel west of the Jordan River (Lepidoptera). Israel J. Ent., 17:23-36.

BENYAMINI, D. 1984. The butterflies of the Sinai Peninsula (Lep., Rhopalocera). Nota Lepid., 7:309-321.

BERNARDI, G. 1972. Note sur la variation de l'armure genitale des Iolana. Bull. Soc. ent. Fr., 77:160-167.

BERNARDI, G. & GILLET, H. 1962. Sur l'existence du Colitis fausta Oliv. dans l'Ennedi (Nord Tschad). Bull. Soc. ent. Fr., 67:130-132.

BETHUNE-BAKER, G. T. 1894. On some Lepidoptera received from the neighbourhood of Alexandria. Trans. ent. Soc. London, 1894:33-51.

BODENHEIMER, F. S. & THEODOR, O. 1929. Ergebnisse der Sinai-Expedition 1927 der Hebräischen Universität, Jerusalem. Leipzig.

BOYD, A. W. 1917. Six months' collecting between Ismailia and el Arish. Bull. Soc. Ent. Egypte., 5:98-117.

CHNEOUR, A. 1954. Macrolépidoptères de la Tunisie. I, II. Rhopalocera, Grypocera. Bull. Soc. Sci. nat. Tunis, 7:207-239.

CLARKE, Sir Cyril A. & LARSEN, T. B. 1986. Speciation problems in the Papilio machaon group of butterflies. Syst. Ent., 11:175-181.

CLARKE, C. A. & SHEPPARD, P. M. 1956. A further report on the genetics of the machaon group of swallowtail butterflies. Evolution, 10:103-112.

CLENCH, H. R. 1979 (1978). The names of certain Holarctic hairstreak genera (Lycaenidae). J. Lepid. Soc., 32: 227-281.

DEBSKY, B. 1919. Quelques observations sur Teracolus fausta, Melitaea didyma et Azanus ubaldus. Bull. Soc. Ent. Egypte, 6:33-44.

DEMAISON, L. 1895. Note sur les Lépidoptères d'Egypte. Bull. Soc. ent. Fr., 1895:lix-lxii.

EIG, A. 1931. Les élements et les groupes phytogéographiques auxilliaires dans la faune Palestinienne. Feddes Repert., Spec. Nov. Reg. Veg. Beihandl., 63:1-201.

EVANS, W. H. 1949. A catalogue of the Hesperiidae of Europe, Asia and Australia in the British Museum (Natural History). London.

EVANS, W. H. 1955. A revision of the genus Tarucus (Lepidoptera; Lycaenidae) of Europe, North Africa and Asia. Entomologist, 88:179-187.

FIELD, J. 1971. Butterflies of the genus Vanessa and the resurrected genera Bassaris and Cynthia. Smithson. Contr. Zool., 84:1-105.

FELTWELL, J. 1981. The large white butterfly; the biology, biochemistry and physiology of Pieris brassicae Linné. Junk, Amsterdam.

GABRIEL, A. G. & CORBET, A. S. 1949. Results of the Armstrong College expedition to Siwa Oasis (Libyan Desert), 1395 (sic!), under the leadership of prof. J. Omer-Cooper. Lepidoptera Rhopalocera. Bull. Soc. Fouad 1er Ent., 33:373-379.

GEIGER, H.-J. & SCHOLL, A. 1982. Pontia daplidice in Südeuropa – eine Gruppe von Zwei Arten. Mitt. schweiz. ent. Ges., 55:107-114.

GOOD, R. 1977. The geography of flowering plants. Longman, London.

GRAVES, P. P. 1905. Notes on collecting in Egypt. Ent. Rec. J. Var., 17:147-150.

GRAVES, P. P. 1907a. Notes on Egyptian and Syrian butterflies. Ent. Rec. J. Var., 19:64-68.

GRAVES, P. P. 1907b. Note on collecting Lepidoptera in Egypt in May and June. Ent. Rec. J. Var., 19:210-211.

GRAVES, P. P. 1911. A contribution to the fauna of Syria. Ent. Rec. J. Var., 22:31-36.

GRAVES, P. P. 1916 [1915]. A list of the butterflies of Egypt with some notes on those of the Sinai. Bull. Soc. Ent. Egypte, 4:135-157.

GRAVES, P. P. 1918. Some new forms of Lycaenidae from Egypt. Entomologist, 51:97-98.

GRAVES, P. P. 1919. Collecting in various places in 1916-1918. Ent. Rec. J. Var., 31:5-8,61-66, 156-157.

HANCOCK, D. L. 1983. Classification of the Papilionidae; a phylogenetic approach. Smithersia, Zimbabwe, 2:1-48.

HAYWARD, K. J. 1925. Notes on Egyptian Lepidoptera observed at the Reservoir, Aswan. Ent. Rec. J. Var., 37:(1)-(5); and 1926, ibid, 38:(5)-(13).

HIGGINS, L. G. 1941. An illustrated catalogue of the Palaearctic Melitaea (Lep.; Rhopalocera). Trans. R. ent. Soc., Lond., 91:175-365.

HERZOG, J. 1911. Mein Schmetterlingsfang in Ägypten und Syrien. Jahrb. ent. Ver. Wien., 21:85-96.

HIGGINS, L. G. 1956. Bemerkungen zu Melitaea und Pararge Formen aus dem mittlerem Osten, veranlasst durch F. Heydemanns kürzlich erschienenden Artikel über Rhopaloceren aus Afghanistan. Z. Wien. ent. Ges., 41:10-15.

HONORÉ, A. 1939. Migration de Precis cebrene Trimen en Egypte (Lep., Rhopalocera). Bull. Soc. Fouad 1er Ent., 22:284-285.

INNES-BEY, W. 1909. Resumé des travaux relatifs en partie la faune de l'Egypte. Bull. Soc. Ent. Egypte, 2:10-16.

INNES-BEY, W. 1909a. Bibliographie des travaux entomologiques relatifs a l'Egypte. Bull Soc. Ent. Egypte, 2: 128-144.

JACKSON, D. J. 1910. Some notes on Egyptian butterflies. Ent. Rec. J. Var., 22:17-19.

JONG, R. de 1974. Systematics and evolution of the Palaearctic Spialia species (Lepidoptera; Hesperiidae). Tijds. voor Ent., 117:225-271.

JONG, R. de 1978. Carcharodus tripolinus Verity, stat. nov. une nouvelle espèce pour la faune d'Europe. Linn. Belgica, 7:117-122.

KLUG, F. & EHRENBERG, C. G. 1829-1845. Symbolae physicae seu icones et descriptiones insectorum ....etc. Berlin.

KNEUCKER, A. 1903. Zoologische Ausbeute einer botanischen Studiereise durch die Sinaihalbinsel im März und April 1902. Verh. zool.-bot. Ges. Wien, 53:575-587.

KUDRNA, O. 1974. Artogeia Verity 1947, gen. rev. for Pieris napi Linnaeus (Lep.; Pier). Ent. Gaz., 25:9-12.

KUDRNA, O. 1977. A revision of the genus Hipparchia Fabricius. Classey, Hampton.

KUDRNA, O. 1986. Butterflies of Europe. Vol. 8. Aspects of the conservation of butterflies in Europe. Wiesbaden.

LARSEN, T. B. 1974. Butterflies of Lebanon. Beirut.

LARSEN, T. B. 1975. Provisional notes on butterfly migration in Europe. Atalanta, 6:62-74.

LARSEN, T. B. 1976. The importance of migration to the butterfly faunae of Lebanon, East Jordan and Egypt. Notul. entomol., 56:73-83.

LARSEN, T. B. 1977a. Notes on the distribution and subspecies of Euchloe charlonia Donzel (Lep.; Pieridae). Atalanta, 8:107-109.

LARSEN, T. B. 1977b. A migration of Anaphaeis aurota F. in Kenya. Dtsch. entomol. Z., 24:429-432.

LARSEN, T. B. 1977c. Il y a trois millénaires et demi que Danaus chrysippus Linné est connu en Haut Egypte. Linn. Belgica, 7:55-58.

LARSEN, T. B. 1977d. The butterflies of eastern Oman and their zoogeographical composition. J. Oman Stud., Sp. Rpt., 1:179-207.

LARSEN, T. B. 1978a. A mixed butterfly migration in West Africa (Benin). Atalanta, 9:191-198.

LARSEN, T. B. 1978b. The first record of Catopsilia florella F. from Lebanon for more than fifty years. Atalanta, 9:253-254.

LARSEN, T. B. 1979. A contemporary view of Francis Walker. Ent. Gaz., 30:156.

LARSEN, T. B. 1980a. The status of Papilio machaon rathjensi Warnecke and its relationship to other Arabian populations. J. Lepid. Soc., 34:365-367.

LARSEN, T. B. 1980b. The butterflies of Dhofar and their zoogeographical composition. J. Oman Stud., Sp. Rpt., 2:153-186.

LARSEN, T. B. 1982a. The butterflies of the Yemen Arab Republic (with a review of the species in the Charaxes viola-group from Arabia and East Africa). Biol. Skr. Dan. Vidensk. Selsk., 23 (3):1-85.

LARSEN, T. B. 1982b. Mass population build-up of Anaphaeis aurota F. in Oman. Atalanta, 13:42-47.

LARSEN, T. B. 1982c. The importance of butterfly migration to the fauna of Arabia. Atalanta, 13:248-259.

LARSEN, T. B. 1983. Insects of Saudi Arabia; Lepidoptera, Rhopalocera (a monograph of the butterflies of Arabia). Fauna of Saudi Arabia, 5:333-478.

LARSEN, T. B. 1984a. Butterflies of Saudi Arabia and its neighbours. Stacey Intl., London.

LARSEN, T. B. 1984b. A tropical migrant butterfly new to the eastern Mediterranean, Junonia orithya here Lang. Atalanta, 15:101-102.

LARSEN, T. B. 1984c. The zoogeographical composition and distribution of Arabian butterflies (Lepidoptera; Rhopalocera). J. Biogeogr., 11:119-158.

LARSEN, T. B. 1985. A mixed butterfly migration in the Nairobi area of Kenya. Atalanta, 16:104-105.

LARSEN, T. B. 1986. Tropical butterflies of the Mediterranean. Nota Lepid., 9:63-77.

LARSEN, T. B. 1987. Biogeographical aspects of Middle Eastern and Arabian Butterflies. Proceedings of the symposium on the fauna and zoogeography of the Middle East. Beihefte zum Tübinger Atlas des vorderen Orients, A (Naturwissenschaften), nr. 28:178-199 (Krupp, F., Schneider, W. & Kinzelbach, R., eds.).

LARSEN, T. B. & LARSEN, K. 1980. The butterflies of Oman. Bartholomew, Edinburgh.

LARSEN, T. B. & NAKAMURA, I. 1983. The butterflies of East Jordan. Ent. Gaz., 34:135-208.

LARSEN, T. B. & PITTAWAY, A. R. 1982. Notes on the ecology, biology and taxonomy of Apharitis acamas (Klug) (Lep.; Lycaenidae). Ent. Gaz., 33:163-168.

MEERMAN, J. & BOOMSMA, T. 1986. Die wenig bekannte Schwalbenschwanz Papilio saharae rathjensi Warnecke 1932. Freilandsbeobachtungen und Zucht. Ent. Z., 96:177-192.

MANDERS, N. 1915. The butterflies of lower Egypt. Ent. Rec. J. Var., 27:60-65.

NAKAMURA, I. 1975. Description of two new species of butterflies (Lepidoptera; Lycaenidae) from the South Sinai. J. Ent. (B), 44:283-295.

NAKAMURA, I. 1977. On the identity of some Lycaenidae species described by Francis Walker (1870) from Egypt and the Sinai. Ent. Gaz., 28:9-14.

NAKAMURA, I. & BENYAMINI, D. 1973. Euchloe falloui Allard (Lepidoptera; Pieridae) in the Sinai Peninsula and the Negev desert. Entomologist, 106:267-268.

PACLT, J. 1949. Nomenclator sive enumeratio critica Lepidopterorum Slovaciae. Ent. Listy., 12:140-155.

PARKER, R. 1983. The butterflies of Cyprus. Ent. Gaz., 34:17-53.

PITTAWAY, A. R. 1979. The butterflies and hawkmoths of eastern Saudi Arabia. Proc. Brit. nat. ent. Hist. Soc., 12:90-101.

PITTAWAY, A. R. 1980. Butterflies (Lepidoptera) of Qatar, April-June, 1979. Ent. Gaz., 31:103-111.

PITTAWAY, A. R. 1985. Lepidoptera Rhopalocera from western Saudi Arabia. Fauna of Saudi Arabia, 7:172-197.

REBEL, H. & DAUB, M. 1909. Lepidoptera aus der Sinaihalbinsel und Ägypten. In KNEUCKER, A. Zoologische Ergebnisse zweier botanischer Studienreisen in Ägypten. Verh. naturwiss. Ver. Karlsruhe., 21 129-146.

RILEY, N. D. 1925. The species usually referred to Cigaritis Boisduval (Lepidoptera; Lycaenidae). Novit. zool., 27:70-95.

ROELL, L. 1959. Frühlingsfahrt nach Ägypten. Ent. Z., 69:205-218.

ROSE, K. & SCHURIAN, K. 1977. Beiträge zur Kenntniss der Rhopaloceren Irans. 10. Neue Lycaeniden-Unterarten aus Südiran. Ent. Z., 87:145-151.

SAYED, M. T., ROSTOM, Z. M. F. & KASHEF, A. H. 1964. Contribution to the insect fauna of some oases in the Egyptian western desert. Bull. Soc. Ent. Egypte, 48: 259-267.

SHAPIRO, A. M. 1980. Mediterranean climate and butterfly migration; an overview of the Californian Fauna. Atalanta, 11:181-188.

SCHURIAN, K. G. & HOFMANN, P. 1982. Die Thersamonia Gruppe. Nachr. ent. Ver. Apollo, Suppl. 2:1-59.

SOMEREN, V. G. L. van 1971. Revisional notes on the African Charaxes; part VII. Bull. Brit. Mus. nat. Hist., (Ent.), 23:193 ff.

STEMPFFER, H. 1969. Contribution a l'étude des Lycaenidae de la faune éthiopiénne. Bull. Soc. ent. France, 74:198-199.

STEPHANIAN, S. 1911. Les insectes ravageurs des rizires en Egypte. Bull. Soc. Ent. Egypte: 84-86.

STOREY, G. 1914. Notes on some Egyptian Lepidoptera. Bull. Soc. Ent. Egypte, 4:73-75.

TÄCKHOLM, V. 1974. Student's flora of Egypt. Cairo University Press, Beirut.

WALKER, F. 1870. A list of the butterflies collected by J. K. Lord esq. in Egypt, along the African shore of the Red Sea, and in Arabia. Entomologist, 5:48-57.

WILLIAMS, C. B. 1929. The seasonal abundance of four common butterflies in Egypt. Bull. Soc. Roy. ent. Egypte, 13:85-92.

WILLIAMS, G. H. 1917. Notes on some Egyptian butterflies. Bull. Soc. Ent. Egypte, 5:91-94.

WILTSHIRE, E. P. 1947. Egyptian Lepidoptera. Some additions and corrections to previous lists. Bull. Soc. Fouad 1er Ent., 31:13-16.

WILTSHIRE, E. P. 1948. The Lepidoptera of the Kingdom of Egypt. Bull. Soc. Fouad 1er Ent., 32:203-294.

WILTSHIRE, E. P. 1949. The Lepidoptera of the Kingdom of Egypt. Bull. Soc. Fouad 1er Ent., 33:381-460.

ZOHARY, M. 1973. Geobotanical foundations of the Middle East. 2 vols. Stuttgart.

Index